Textile Science and Clothing Technology

Series Editor

Subramanian Senthilkannan Muthu, SgT Group & API, Hong Kong, Kowloon, Hong Kong

This series aims to broadly cover all the aspects related to textiles science and technology and clothing science and technology. Below are the areas fall under the aims and scope of this series, but not limited to: Production and properties of various natural and synthetic fibres; Production and properties of different yarns, fabrics and apparels; Manufacturing aspects of textiles and clothing; Modelling and Simulation aspects related to textiles and clothing; Production and properties of Nonwovens; Evaluation/testing of various properties of textiles and clothing products; Supply chain management of textiles and clothing; Aspects related to Clothing Science such as comfort; Functional aspects and evaluation of textiles; Textile biomaterials and bioengineering; Nano, micro, smart, sport and intelligent textiles; Various aspects of industrial and technical applications of textiles and clothing; Apparel manufacturing and engineering; New developments and applications pertaining to textiles and clothing materials and their manufacturing methods; Textile design aspects; Sustainable fashion and textiles; Green Textiles and Eco-Fashion; Sustainability aspects of textiles and clothing; Environmental assessments of textiles and clothing supply chain; Green Composites; Sustainable Luxury and Sustainable Consumption; Waste Management in Textiles; Sustainability Standards and Green labels; Social and Economic Sustainability of Textiles and Clothing.

More information about this series at http://www.springer.com/series/13111

Subramanian Senthilkannan Muthu
Editor

Recycled Polyester

Manufacturing, Properties, Test Methods, and Identification

Springer

Editor
Subramanian Senthilkannan Muthu
SgT Group & API
Hong Kong, Kowloon, Hong Kong

ISSN 2197-9863 ISSN 2197-9871 (electronic)
Textile Science and Clothing Technology
ISBN 978-981-32-9558-2 ISBN 978-981-32-9559-9 (eBook)
https://doi.org/10.1007/978-981-32-9559-9

This Springer imprint is published by the registered company Springer Nature Singapore Pte Ltd.
The registered company address is: 152 Beach Road, #21-01/04 Gateway East, Singapore 189721, Singapore

Contents

Properties of Recycled Polyester

P. Senthil Kumar and G. Janet Joshiba

Abstract Recycled polyesters are basically derived from the used and waste polyester bottles, old garments and industrial polyester waste, subsequently these recycled polyesters are prepared using two various methodologies such as Chemical and mechanical recycling methods. There is no much difference between virgin polyesters and recycled polyesters in terms of high strength, durability, versatility, and performance. As the Co_2 emission is less in the production of recycled polyesters, it does not pose any threat to the environment. This polyester also reduces the wastage of petroleum fuels which is a major constituent in the production of polyesters, Moreover, it leads to a reduction of overall carbon foot print of the polyester production. The recycled polyesters possess outstanding and enhanced mechanical and thermal properties. This chapter depicts the various fantabulous properties and characteristics of recycled polyesters in detail.

Keywords Recycled polyesters · Mechanical · Thermal · Durability · Strength

1 Introduction

The augmenting surge of industrialization and population has contributed to the elevated disposal of non-degradable plastic waste. Plastic pollution is one of the major alarming issues faced all over the universe [1]. Due to the toxic and non-degradable nature of plastics, it is considered as one of the major pollutant damaging the living organisms and ecosystem. Plastic pollution has elevated its harmful consequences across the marginal and seaside ecosystem universally, it is reported that around 4.8–12.7 million tons of plastics are released into the marine environment [2]. Several researchers have investigated the negative impacts of plastics on aquatic and earth bound ecosystem to reduce its hazardous environmental ill effects. Furthermore,

P. Senthil Kumar · G. Janet Joshiba
Department of Chemical Engineering, SSN College of Engineering, Chennai 603110, India

P. Senthil Kumar (✉)
SSN-Centre for Radiation, Environmental Science and Technology (SSN-CREST), SSN College of Engineering, Chennai 603110, India
e-mail: senthilchem8582@gmail.com

S. S. Muthu (ed.), *Recycled Polyester*, Textile Science and Clothing Technology,
https://doi.org/10.1007/978-981-32-9559-9_1

many studies have conducted to monitor the disposal, effect, and transfiguration of plastics on the surroundings [1]. Plastics are one of the leading pollutants present in higher amount in the marine environment; furthermore, this plastic pollution has become one of the greatest hindrances for the tourist sectors of the coastal regions [2]. Plastics, when disposed of, are not only a worry (or opportunity) for the industry but at the same time is a worry for the legislature that needs to deal with the Municipal Strong Waste [3]. Although plastic has numerous profitable uses in society, it is preferred only for single time usage or as expendable plastic. Around the globe, nearly one million plastic drinking water bottles are utilized every moment, at the same time up to 5 trillion single usage plastic packs are utilized worldwide consistently (www.unenvironment.org). The high strength, less density, wear resistance, abrasion resistance and corrosion resistance are some of the unique properties of the plastic which makes it available for versatile application [4]. The plastic waste found in the land background is due to the floods, over discharge of drainage, an increase of dumpsites, improperly maintained landfills and sudden natural disasters. The plastic pollution in the aquatic systems is due to the discharge of wastes from ships, dumping of non-degradable wastes under the sea and other commercial sources [5]. The waste plastic released from the industries, commercial and domestic sources are distinguished into macro plastics and micro plastics based upon the size of the plastic. Usually, the macroplastics are larger plastic compounds which are of the size range greater than 25 mm, in addition, consumption of these macroplastics causes damage to the living organisms present in the aquatic ecosystem. Microplasticsarethe plastic compounds which are of the size range less than 5 mm, moreover, it is primarily utilized as raw materials in the cosmetic industry, pharmaceutical industry, and textile industry. As these plastics are complex to degrade, they easily enter the food chain of the living organisms which further results in bio accumulation and bio magnification [2, 6]. In order to combat all the deleterious effects of the plastics, effective measures for plastic waste management have been carried out. As the plastic stake a large span of time for degradation the complete degradation of plastics is not possible. In addition, the disposal of plastics through pyrolysis and landfills also aggravate several harmful consequences in the environment [7]. Further, to reduce the ill effects of the waste generated, the industries are insisted to follow environmental friendly manufacturing practices. Many environmental and governmental organizations have made attempts to publicize the green industries and its application to conceive consciousness about the difficulty in waste disposal. The 3R concept was seemed to be one of the efficient ways of Reducing and managing industrial waste. The recycling method is a classical technique used for a longer time period to reduce the environmental ill effects caused due to the accumulation of hazardous nondegradable waste products [8].

The textile business is of incredible significance to the economies of each nation. In addition, it occupies a major place in elevating the venture, income, business, and exchange. In 2015, the span of the worldwide clothing market came to USD 1685 billion [9]. Textile waste is seemed to be one of the leading industry utilizing polyester fibers for garment manufacturing. While washing the garments, the polyester fibers release microplastics as the major contaminants into the aquatic systems. It is reported that the laundering of about 5 kg of polyester materials emits about 6,000,000 microplastic fibers into the nearby marine sources [2]. In order to compensate all the deleterious effects of the polyesters on the environment, the polyester recycling is considered as one of the promising approaches. This chapter elaborates about the recycled polyesters and also the various unique properties of the recycled polyesters.

2 Polyester

Polyester fiber is basically prepared from petroleum which is a high demand non-renewable fuel source. The extraction of petroleum from crude oil creates a negative impact on the wellness of the ecosystem. In spite of the fact that it is a typical suspicion that materials produced using common materials, for example, cotton and silk are more ecologically cordial than materials produced using man-made materials proven to be wrong in the case of polyester fibers. When keenly examining the entire life cycle assessment of the polyester fibers, it is seemed to be eco-friendly than the other natural fibers. Polyester has become one of the leading fiber categories among the worldwide textile industries (www.tedresearch.net). Polyethylene terephthalate (PET) polyester is one of the widely utilized thermoplastic polyesters and it is commonly known as "Polyesters" (Fig. 1).

The PET occurs in amorphous and semi-crystalline state of the material. In addition, it is well known for its unique characteristics such as hardness, strength, flexibility, rigidity, etc., [10]. Poly(ethylene terephthalate) (PET) is a standout amongst the most widely reused polymeric materials. The fundamental inspiration for the across the board recycling of PET is its broad use in plastic bundling applications, particularly in the refreshment industry [3]. This material is complex to degrade and it possesses a long shelf life of 35–42 years to degrade. Initially, the virgin and recycled polyesters with 75D and 150D were obtained from Polygenta ventures, Nasik and Manjushree enterprises, Ichalkaranji (www.textilemates.com). The three predominant polyesters available in the market for industrial and commercial applications are Polycarbonate, Polyethylene terephthalate, and polybutylene terephthalate. Polyesters are basically synthesized using condensation reaction, in which a dialcohol and diacid are reacted with each other to form the polyesters with the release of water. The polyesters are prepared in the same methodology in which the polyamides and nylons are synthesized [10].

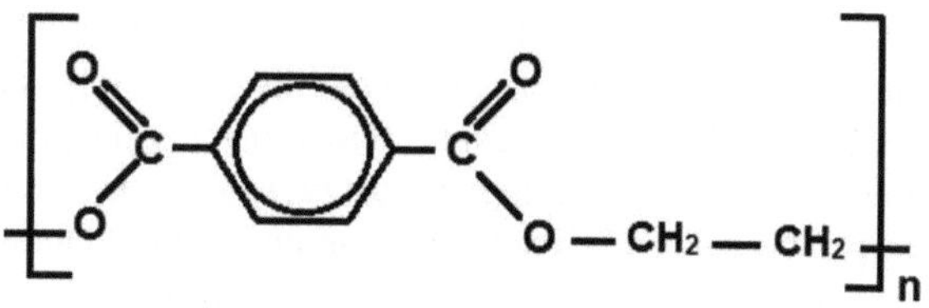

Fig. 1 Chemical structure of polyesters

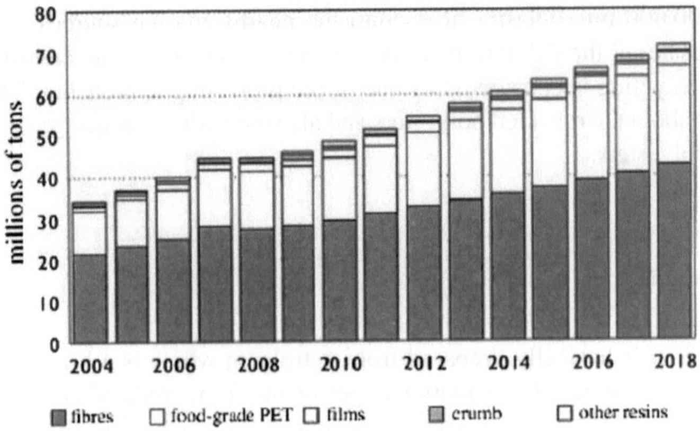

Fig. 2 Graphical representation of the global expenditure of PET [26]

Polyethylene terephthalate is one of the remarkable polyester materials utilized in various industrial, commercial and domestic applications. It remains as one of the major raw material in many appliances such as textiles, beverages, food packaging, water bottles, kitchen containers, hearing aids, thin films, sheets, sports items, construction materials, wrapping materials, etc., Out of the various PET material used the bottle grade PET is one of the widely used polymer material in the current scenario. Majority of the food packaging, water bottles and beverages holding containers are made out of the PET materials, Moreover, these bottles comprise about 83–84% of worldwide PET material demand. It is reported that the anticipated interest for PET bundling materials is determined to achieve 20 million tons by 2019 with a yearly increment of 4.6%. This elevated utilization of the PET substantially increased the emission of large amounts of PET waste into the environment [7]. The total worldwide consumption of PET in various forms for different applications is depicted in Fig. 2. Due to their unique properties, they are utilized in various applications such as Water bottles, Food packings, Medications, Textiles, Carpets, Audio cassettes, insulation tapes, automobile parts, electrical hardware compounds, and fabrics.

3 Need for Recycled Polyester

The textile industry is one of the major industry required for maintaining the lifestyle of human beings, therefore the requisition of textile fiber is augmenting globally. In this case, two major fibers such as cotton and polyester command the overall textile industries. Within a short span of fifteen years, the polyester fibers have overwhelmed the cotton fiber as the single most prevalent material [11]. The pollution caused due to plastic creates an extensive dispute among mankind; furthermore, the toxic and non-degradable nature of the plastic materials complicates the ecosystem [8]. The extended utilization of polyester has increased the emission of polyester waste into

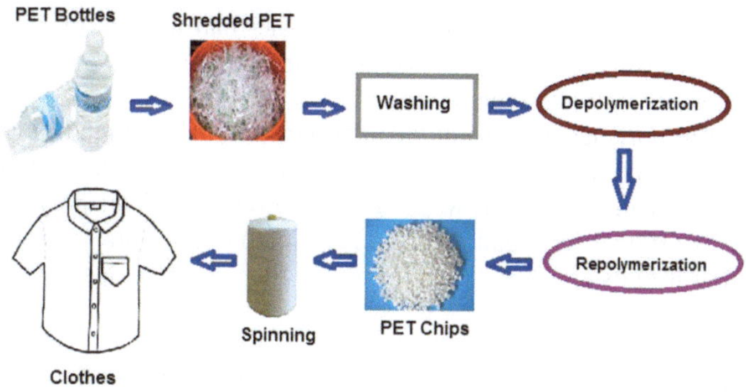

Fig. 3 Process of polyester recycling

the environment. These materials do not pose any direct damage to the living organisms and the ecosystem, but their nonbiodegradable nature remains as one of the biggest threat worldwide resulting in the elevation of polyester solid waste. In addition, their greatest defiance against the biological degradation and environmental factors pulls the polyesters into the list of major harmful pollutants. The remediation of polyester waste can be effectively done by two approaches such as incineration and recycling. As the incineration process emits noxious gases into the environment which are capable of causing adverse health disorders and damaging the ecosystem. Uncertainty, the recycling approach will be chosen as an effective method for managing polyester waste. The recycling of polyesters reduces the higher expenditure involved in the production of virgin fibers. In addition, it also paves the way in conservation of energy sources [7]. The process of recycling of polyesters is depicted in the Fig. 3.

4 Identification of Plastic and Its Recycling Code

Before entering into the recycling process, there is frequently a need to distinguish which specific plastic material has been utilized for the guaranteed item. Most purchasers perceive the kinds of plastics by the numerical coding framework made by the Society of the Plastics Industry (SPI) in the late 1980s. The sort of plastic that the compartment is made out of is obviously stepped onto it. The image of a triangle framed by three arrow mark is the common symbol of the type of plastic; there is likewise a number inside and a letter code underneath. Based on the numbers inside the triangle the plastics are distinguished:

(1) **Poly(ethylene terephthalate) (PETE or PET)**
PET is one of the major polymer used in the food packaging industries, water bottles, and construction industries. The recycled PET in the form of fibers is used in the pillow manufacturing and carpet industry. Majority of the recycled

PET is used in the textile industry and another half amount of recycled PET again enters into the water bottle manufacturing sector.

(2) **High-density Polyethylene (HDPE)**
The HDPE polymers are one of the important polymers which has vital fundamental application in the market and it is majorly used in milk jugs, laundry detergents, shampoo bottles, washing and shower soaps. This polymer is colorless in its original form and it can be recycled into any color and shape. After recycling it finds applications in trashcans, funnels, stumbles, window boxes, etc.,

(3) **Polyvinyl chloride**
This polymer is mainly used in food packing sectors as trays, food foils to wrap and tetra packs.

(4) **Low-density polyethylene (LDPE)**
This LDPE polymer is mainly used in shopping bags, crushed bottles, highly resistant sacks, and wrappings. It is the same as HDPE in its chemical form.

(5) **Polypropylene (PP)**
This polypropylene polymer is also one of the widely used polymers in the environment. It has several applications in the environment such as furniture, refrigerated fillings, toys, luggage and lining and the external border of the cars.

(6) **Polystyrene**
This polystyrene polymer is majorly utilized in the toys, hard packing, refrigerator trays, cosmetic bags, audio cassettes, CD cases, vending cups, and cosmetic bags.

(7) **Other uncoded plastics**
Plastic purchaser merchandise not recognized by code numbers are not normally gathered. These plastics used in a large number of different items basically don't fit into the numbering framework that distinguishes plastics utilized by the consumers.

5 Recycled Polyesters

The recycled polyesters are developed from the used plastic bottles and Polyethylene terephthalate (PET) [11]. The recycled polyesters were initially launched in 1940s The recycling of PET bottles is divided into two major categories such as open loop and closed loop recycling [7]. Recycling is seemed to be the most practical way to decrease the impact caused by the PET solid waste. It was asserted that in 1994 more than 500 million plastic compartments were changed over 45 million pounds of polyester filaments for floor covering and home outfitting. The polyester is claimed to be one of the most recyclable polymers universally. A portion of the items in which reused polyester is utilized as strands, films, froth, sheeting, food packaging and water bottles (www.textilemates.com). The garment industry named H&M has contributed the greatest part in using the waste PET for the textile production. In 2012, this firm has utilized around 7.9 million plastic bottles which are subjected to landfill treatment (Ethically sustained WordPress). During the production of PET fibers the bottles are collected in a bulk quantity and it is subjected to a series of conversion process such as separation, purification, grinding and drying. The fibers obtained from the recycling process are utilized majorly in the textiles, blankets, carpets and other domestic appliances. Some of the studies concluded that the recycling rate of PET bottles has elevated from 13 million tons in 2018 to 15 million tons in 2020. Some of the beneficial features which make the recycled polyesters as an outstanding material are as follows:

- Reduces the consumption of petroleum for polyester production
- Controls the solid waste aggregation
- Decreases the non- degradable PET bottles dumped in the landfills
- Lessen the harmful consequences of plastic on the environment
- Minimizes the wastage
- The recycled polyesters are less harmful than the virgin polyesters
- Minimizes the emission of greenhouse gases and ozone depletion
- Decreases the usage of harmful nondegradable chemicals used in the production of virgin polyesters.

6 Methodology for Polyester Recycling

The polyester is naturally nondegradable material and also it cannot be degraded by any microorganisms. Moreover, it requires expensive technologies for its decomposition and processing [12]. The recycling of the PET polyester waste is seemed to be the best option to reduce the deleterious effects of PET waste [12]. There are three particular ways to deal with the reusing of post-purchaser plastic bundling materials. The bundling may: (1) be reused legitimately; (2) experience physical reprocessing, for instance crushing, dissolving and transforming; (3) be exposed to concoction treatment whereby its parts are disconnected and reprocessed for use in production [3]. The recycling of the polyesters materials is of two types they are primary recycling and secondary recycling. The primary recycling of polyesters is one of the most well-known methods because of its simplicity and feasibility. This procedure alludes to the reuse of items in their unique structure. The main disadvantage in this method is the presence of an utmost on the number of cycles for every material. The secondary recycling technique is the conversion of the polymers into other products by heating. The secondary recycling is of two types such as mechanical and chemical recycling method. In the mechanical recycling the polyester waste is converted into recycled fibers through a series of conversion steps such as cutting, shredding, purification, pellet formation, spinning, and fiber production. The complete process of conversion of PET bottles into fibers is depicted in Fig. 4.

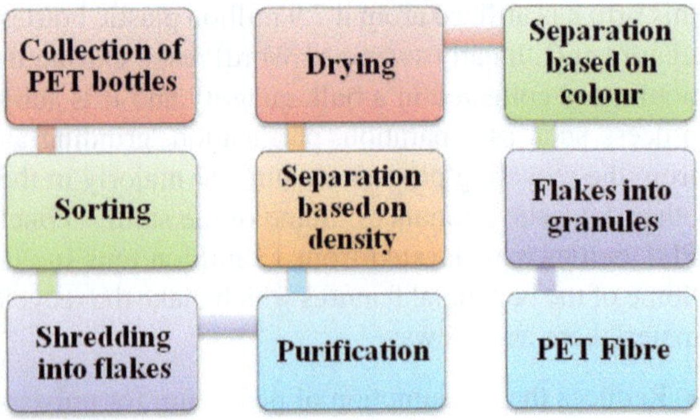

Fig. 4 Conversion of PET to fibre

6.1 Re-Utilization

The reuse is seemed to be the primary and less effective method of recycling the polyesters. The waste polyesters are separated and segregated according to their capability to reuse. The main drawback behind the reutilization chamber is the ingestion of contaminants in the containers.

6.2 Mechanical and Thermal Reprocessing

It is also known as the secondary reprocessing of polyesters. In the mechanical and thermal reprocessing, the polymer materials are subjected to a series of conversion process such as collection, purification, grinding, shredding, melting, molding, spinning, pelletization. In this methods, the virgin polyesters are recreated into other recycled polyesters.

6.3 Chemical Reprocessing

In this method, the virgin polymers are reprocessed chemically and converted into monomers. This process includes a series of conversion steps such as collection, segregation, and degradation. The partial processing is considered to be the economical method, whereas when the polymers are chemically reprocessed completely it becomes quite expensive and requires high-cost raw materials. Chemical recycling is the conversion of multiple branched polymers to monomers through chemical reactions. This type of recycling alters the chemical structure of the polymers and it decomposes the polymers into monomers through various chemical reactions such as Glycolysis, Pyrolysis, Cracking, Hydrolysis, Methanolysis, Photodegradation and Thermal cracking [4, 13, 14].

6.4 *Factors Influencing the Polyester Reprocessing*

Polyester recycling depends on some of the physical and chemical features such as.

6.5 *Colorants*

The water bottles and other colored polymer compounds release colors and pigments during the recycling process. The effective segregation and purification recycling procedure lower the colored compounds.

6.6 *Contaminants*

The contaminants present in the polyester flakes are one of the important parameters depending on which the polyester is recycled. The type and intensity of the contaminant in the polyester have a strong impact on the physical and chemical properties of the recycled polyesters. Some of the contaminants release toxic acids during the reprocessing steps which remain as a major hindrance in the polyester recycling. The acetaldehyde is one of the toxic compound released during the PET recycling, in addition, the acetaldehyde enters the food chain easily and affects the environment. The adhesives and poly vinyl chloride are some of the contaminants present in the polyesters [15].

6.7 *Water*

The presence of water in the polyester during the recycling process reduces the weighted average of the material. The moisture content should be always below 0.02% in the polyester material to be recycled. The presence of a higher amount of water is due to the overflow during the washing process and improper drying process [12].

7 Properties of Recycled Polyesters

The polyesters are generally developed to be hard, soft, frail, flexible and strong materials. The thermoplastic polyesters are identical with the properties of nylon 6 and nylon 66. The exception is its high dimensional stability and less water absorption capacity. When these polyesters are coated with glass fibers, several features such as strength, strength to weight ration, mechanical properties, and other defiances are enhanced in the polyester materials [10]. The polyesters are well known for their unique properties such as:

- Amorphous structure
- High strength
- High heat resistance
- Flame retardancy
- Dimensional stability
- High impact resistance
- Wear resistance
- Abrasion resistance
- Acoustics absorption
- High tensile strength
- Thermal insulation
- High mechanical strength
- Chemical resistance
- Resistance against UV radiation.

The recycled polyesters are seemed to possess unique outstanding mechanical, thermal insulation, tensile properties, wear resistance, tear resistance, etc., Several studies have been carried out on the properties of the recycled polyesters. The recycled polyesters are represented by some of its characteristics such as color, viscosity, molecular weight, functional groups and thermal properties [8]. The various outstanding properties of the recycled polyesters are explained in this chapter elaborately.

7.1 Physical Properties

The recycled polyesters have remarkable physical properties such as bending length, moisture regain, abrasion resistance, tenacity and crease recovery angle. The recycled polyesters possess enhanced moisture regaining capacity than the virgin polyesters because of the significant molecular arrangement. The virgin polyesters possess higher bending length when compared to the recycled polyesters because of the lessened crystallinity in the recycled polyesters. In addition, the higher tenacity also enhances the mechanical strength of the recycled polyesters (www.textilemates.com). The results of the physical and molecular examination concluded that the virgin polyesters are not same as the recycled polyesters. The virgin polyesters seemed to possess higher molecular weights, higher viscosity and less melt flow index whereas the recycled polyesters possess less molecular weight, less viscosity because of the impact of higher temperature and shear caused during the recycling process of the PET.

7.2 Mechanical Properties

The mechanical properties of the polyesters are estimated in the form of tensile strength, flexibility, randomness and defects present in the yarn and fabric [16]. In the study conducted by Jo et al. the mechanical properties of the recycled polyesters are studied and their impact in the polymer concretes has been studied. Due to the effective mechanical properties of the recycled polyesters, it is utilized in the construction sectors to enhance the strength of the concretes. The studies concluded that the recycled polyesters possess a compressive strength of about 73.7 MPa, Tensile strength of about 7.85 MPa, Flexural strength of about 22.4 MPa and elasticity of about 27.9 MPa at a time period of 7 days. A strong interrelationship is seen in between the compressive strength and other physical properties of the polyesters such as elasticity, tensile strength, and flexibility. The application of the recycled polyesters on the concretes have some major advantages such as conservation of energy resources, minimum expenditure, reduction of solid waste [17]. Saeed et al. have investigated the properties of recycled polyester fibers. In this research work, the virgin hyperbranched polyester is fabricated and combined with recycled polyester fibers, consecutively; both the virgin and recycled polyesters are subjected to spinning and thermal drawing process. The characterization results concluded that the mechanical properties of the fibers are seemed to be higher than the virgin polyesters. The enhanced mechanical properties of the recycled polyesters are due to the distributed arrangement of the fibers along the axis [16]. Jo et al. have studied the mechanical properties of polymer concretes made out of recycled polyester materials. In this study, the acid resistance, stress resistance and mechanical strength of the concrete prepared from the recycled polyesters are studied. The mechanical properties of the recycled polyesters are seemed to be effective than the normal concrete material [18]. Lin et al. have implemented the recycled polyethylene terephthalate waste as short fibres as hardening cementitious composites. The results seemed to be convincing and it declared that this recycled PET materials can be potentially used in the construction industry and it is also seemed to be economically feasible and eco-friendly technique [19]. The Post- consumer PET materials exhibit lower intrinsic velocity after thermal and mechanical reprocessing system because of the impacts created by hydrolytic and thermal degradation. In addition, the hydrolytic degradation lowers the tensile and mechanical properties. To combat this issue, the PET materials are extruded with some chain elongation such as bis-iminocarbonate, diphenyl carbonate, bis-2-oxazolines, diphenyl oxalate, pyromellitic dianhydride (PMDA) and diisocyanate [20]. Yuksekkaya et al. have investigated the comparative analysis of the mechanical properties of the virgin and recycled polyesters. The results concluded that the recycled polyester fibres exerted enhanced mechanical properties when it is compared with the virgin polyesters. Also, this study clearly Inferred that the recycled polyester fibres have higher tensile strength and it is highly suited in the manufacturing of sweatshirts, T-shirts, etc., [16].

7.3 Thermal Properties

The thermal properties and fire retardancy properties of the recycled polyesters encapsulated with magnesium hydroxide nanocomposites is investigated by Naugib. The thermal properties are generally studies using thermal gravimetric analysis. The results concluded that the nanocomposites enclosed recycled polyesters showed great thermal properties because of the dehydration of $Mg(OH)_2$ and inhibition of CaO layers [21]. The PET polymers are generally flammable materials and they do not possess great fire retardancy characteristics. The fire retardancy characteristics of the PET polymer fibers are enhanced by combining it with synthetic halogen composed fire retardants [21]. Generally, the polyesters are prone to hydrolytic degradation which affects the fire resistance properties of the fibers. To combat these defects the processing of recycled polyesters are performed using the polycondensation method. Delva et al. have utilized the Brominated polystyrenics combined with the recycled polybutylene terephthalate for enhancing its fire reatardancy [22]. Naugib investigated the thermal stability and fire resistance characteristics of the recycled polyesters encapsulated with $Mg(OH)_2$. In this work, the recycled PET polyester is combined with the synthesized $Mg(OH)_2$ nanocomposites through in situ polymerization technique and tested for their flame resistance properties. The results concluded that this recycled polyesters possess great fire resisting capacity due to its prolonged time for ignition, mass loss rate and smoke rate [21].

7.4 Sound Absorption Properties

The less specific gravity and feasibility are the two main factors contributing to the utilization of the textile fibers for effective sound absorption properties. In common, the materials which absorb sound are majorly categorized into three types such as porous, resonator and panel. Lee and Joo have investigated about the sound absorption characteristics of the recycled polyester fibers. They determined the interrelationship between the properties of the fiber and sound absorption coefficient. In this experimental work, the sound absorption capacity of the recycled polyesters is compared with the conventional woven materials such as rock wool and glass wool. [23]. Patnaik et al. have investigated the thermal properties of the recycled polyester fibers and wool fibers. Various proportions of the woolen and recycled polyester fibers are examined for the properties such as fire resistance, sound insulation, thermal insulation, and moisture retardancy. The recycled polyesters combined with the woolen fibers showed an effective sound absorption capacity of about 70% in the frequency range of 50–5700 Hz [24]. The recycled polyesters are preferred more than the conventional virgin fibers due to some of the factors such as less expense, easy handling and outstanding properties [23].

7.5 *Environmental properties*

The degradation of plastic materials results in the release of several polyesters such as polylactic acid, poly-3-hydroxyalkanoate, polybutylene succinate and polycaprolactone. Many of the polyesters found in our environment are capable of degrading under the presence of some enzymes produced by microorganisms. The enzymatic degradation of polyesters occurs through chemical processes such as hydrolysis, depolymerisation and conversion into CO_2 and H_2O. One of the main limitation in the enzymatic degradation of the polyesters is the less occurrence of these polyester degrading bacteria [25].

8 Conclusion and Future Perspectives

This chapter elaborates about the recycled properties and their unique properties such as mechanical, thermal and sound absorption properties. There is a tremendous chance to demonstrate the reusing procedure and increment consistency of it, for that inline atomic weight and carboxyl end-bunch content assurance would help as a delicate sensor. The rheological studies should be developed for effective segregation of the polyesters according to the color and type of material, in addition, this study helps in understanding the nucleation process. The effective segregation of the polyesters depending on the color enhances the recycling capability of the polyesters. The implementation of more sophisticated instrumentation for characterization studies helps in studying the properties of the recycled polyesters. The special sensors should be implemented to determine the contaminants, moisture content and coloring agents which reduce the effectiveness of the mechanical and thermal properties of the polyesters.

References

1. Chae Y, An YJ (2018) Current research trends on plastic pollution and ecological impacts on the soil ecosystem: a review. Environ Pollut 240:387–395
2. Wang M, He Y, Sen B (2019) Research and management of plastic pollution in coastal environments of China. Environ Pollut 248:898–905
3. Nikles DE, Farahat MS (2005) New motivation for the depolymerization products derived from poly(ethylene terephthalate) (PET) waste: a review. Macromol Mater Eng 290:13–30
4. Grigore ME (2017) Methods of recycling, properties and applications of recycled thermoplastic polymers. Recycling 2:24
5. Liubartseva S, Coppini G, Lecci R (2019) Are mediterranean marine protected areas sheltered from plastic pollution? Mar Pollut Bull 140:579–587
6. Stafford R, Jones PJS (2019) Viewpoint—Ocean plastic pollution: a convenient but distracting truth? Marine policy 103:187–191

7. Sarioglu. E, Kaynak HK (2018) PET bottle recycling for sustainable textiles. In: Camlibel NO (ed) Polyester—production, characterization and innovative applications. Intech open science, New York
8. Vadicherla T, Saravanan D, Muthu SSK (2015) Polyester recycling—Technologies. In: Muthu SSK (ed) Characterisation, and applications, environmental implications of recycling and recycled products, pp 149–165
9. Filho WL, Ellams D, Han S, Tyler D, Boiten V, Paço A, Moora H, Balogun AL (2019) A review of the socio-economic advantages of textile recycling. J Clean Prod 218:10–20
10. McKeen LW (2010) Polyesters. In: McKeen LW (ed) Fatigue and tribological properties of plastics and elastomers, 2nd edn, pp 99–147
11. Fletcher K (2012) Sustainable fashion and textiles: design journeys. Routledge, London
12. Awaja F, Pavel D (2005) Recycling of PET. Eur Polym J 41:453–1477
13. Upasani PS, Jain AK, Save N, Agarwal US, Kelkar AK (2011) Chemical recycling of PET flakes into yarn. J Appl Polym Sci 123:520–525
14. Karayannidis GP, Achilias DS (2007) Chemical recycling of poly (ethylene terephthalate). Macromol Mate Eng 292:128–146
15. Giannotta G, Po' R, Cardi N, Tampellini E, Occhiello E, Garbassi F, Nicolai L (1994) Processing effects on poly(ethylene terephthalate) from bottle scraps. Polym Eng Sci 34:1219–1223
16. Yuksekkaya ME, Celep G, Dogan G, Tercan M, Urhan B (2016) A comparative study of physical properties of yarns and fabrics produced from virgin and recycled fibers. J Eng Fibers Fabr 11
17. Jo BW, Park SK, Kim CK (2006) Mechanical properties of polyester polymer concrete using recycled polyethylene terephthalate. ACI Struct J 103
18. Jo BW, Park SK, Park JC (2008) Mechanical properties of polymer concrete made with recycled PET and recycled concrete aggregates. Constr Build Mater 22:2281–2291
19. Lina X, Yu J, Lia H, Lam JYK, Shih K, Sham IML, Leung CKY (2018) Recycling polyethylene terephthalate wastes as short fibers in strain–hardening cementitious composites (SHCC). J Hazard Mater 357:40–52
20. Lei Y, Wu Q, Clemons CM, Guo W (2009) Phase structure and properties of poly(ethylene terephthalate)/high-density polyethylene based on recycled materials. J Appl Polym Sci 113:1710–1719
21. Naguib HM (2018) Environmental-friendly recycled Polyester/Mg(OH)2 nanocomposite: fire-retardancy and thermal stability. Polym Testing 72:308–314
22. Delva L, Hubo S, Cardon L, Ragaert K (2018) On the role of flame retardants in mechanical recycling of solid plastic Waste. Waste Manage 82:198–206
23. Lee Y, Joo C (2003) Sound absorption properties of recycled polyester fibrous assembly absorbers. AUTEX Res J 3
24. Patnaik A, Mvubu M, Muniyasamy S, Botha A, Anandjiwala RD (2015) Thermal and sound insulation materials from waste wool and recycled polyester fibers and their biodegradation studies. Energy Build 92:161–169
25. Wang X, Chen J, Tang X, Wang J, Zhu L, Zhang W, Wang H, Li Y, Zhang Q (2019) Biodegradation mechanism of polyesters by hydrolase from Rhodopseudomonas palustris: an in silico approach. Chemosphere 231:126–133
26. Aizenshtein EM (2010) Still no crisis for polyester fibres. Fibre Chem 42
27. http://www.tedresearch.net/media/files/Polyester_Recycling.pdf
28. https://www.unenvironment.org/interactive/beat-plastic-pollution/
29. https://www.textilemates.com/virgin-recycled-polyester/
30. https://ethicallysustained.wordpress.com/2013/05/08/polyester-and-recycled-polyester/

Effects of Gamma Radiation on the Physicochemical Properties of Polyester Resin and Its Use in Composite Materials

Gonzalo Martínez-Barrera, Angel Martínez-López, Enrique Vigueras-Santiago and Miguel Martínez-López

Abstract In this chapter is discussed the importance of using gamma radiation as a novel technology for recycling of polyester resin, and their applications in composite materials. In the introduction section, some aspects of the environmental problems related with waste polyester resins are discussed. As it is known, novel strategies are required for diminishing the environmental pollution. They are based in no more consumption of non-renewable sources, more use of waste and recycled materials, and the manufacturing of more environmental friendly products, with improved physicochemical properties. The next section is focused on the methods for recycling or reusing of polyester resins. It is well known that mechanical, chemical, and thermal are the most current recycling processes. The mechanical follows the shredding and grinding processes and a method for separation. This has been used for to obtain thermostable polymers, which are used as fillers in different materials. The others processes also are discussed. In the second section, some concepts of gamma radiation are discussed. Such electromagnetic energy, has been used during decades for modifications of the physicochemical properties of different materials. In the case of polymeric materials, the physicochemical changes happen through the processes: scission or cross-linking of polymer chains, and grafting. Each process is produced according to the physicochemical properties of the polymers and the applied dose rate. In the next section, the effects of gamma radiation in polyester resin are mentioned. The polyester resin is usually in the liquid state; once the reaction temperature is reached, this changes from liquid to gel state. Is to say, the gamma radiation produce cross-linking of polymers chains and complete the polymerization process. Gamma radiation shows several advantages, for example it does not require any activation energy for its initiation; and the final reaction can be controlled. The total polymerization depends of the gamma radiation dose and the type of polymer.

G. Martínez-Barrera (✉) · E. Vigueras-Santiago · M. Martínez-López
Laboratorio de Investigación y Desarrollo de Materiales Avanzados (LIDMA), Facultad de Química, Universidad Autónoma del Estado de México, Km. 12 de la Carretera Toluca-Atlacomulco, San Cayetano 50200, Mexico
e-mail: gonzomartinez02@yahoo.com.mx

A. Martínez-López
Doctorado en Ciencia de Materiales, Facultad de Química, Universidad Autónoma del Estado de México, Paseo Colón Esquina Paseo Tollocan S/N, Toluca 50000, Mexico

S. S. Muthu (ed.), *Recycled Polyester*, Textile Science and Clothing Technology,
https://doi.org/10.1007/978-981-32-9559-9_2

In the final section, some studies about gamma radiation as modifier of composites elaborated with polyester resins and different polymers, are discussed. The polymers act as reinforcements or fillers. The irradiated composites show high degree of cross-linking and morphological changes on their surface as well as high improvements on the physical and mechanical properties.

Keywords Polyester resin · Recycling · Gamma radiation · Physicochemical properties

1 Introduction

Novel strategies for diminishing the environmental pollution have been proposed, which covers different environmental problems. Within main objectives are those related to no more consumption of non-renewable sources. As it is known, raw materials extraction and their industrial processes for manufacturing of products provoke contamination. Such circumstances have been factors for to use waste and recycled materials and promote the manufacturing of more environmental friendly products.

Several activities have been designed and followed for to counter the environmental damages, caused by the manufacturing of products. These includes: (a) promote more use of waste materials, mainly those generated as by-products of certain specific process; (b) increase the sustainability through the use of recycled materials, and to avoid the use of raw materials, and (c) to have more durable materials, with improved physicochemical properties; such action allow use less quantities of raw materials [1, 2].

Different polymer artifacts, after their useful life in several activities (sports, household, automotive and aeronautical), cause environmental deterioration. Their degradation propitiates the formation of non-biodegradable particles that flow through the environmental [3].

Nowadays, the choice of materials depends on their properties and costs, as well as some characteristics related to environment and sustainability aspects. Moreover, if they come from the industry, and are considered as wastes. On the last decades, novel methodologies involving sustainability aspects have been adapted to waste and recycled materials. Different materials have been recycled; they include metals, glass or thermoplastics. Nevertheless, recycling of polymer blends, ceramic + polymer or metal + polymer composites, has been difficult, due mainly to their composition [4].

In many situations, there are difficulties for reusing thermostable polymers, due to their characteristics, thus recycling of them is necessary. One of the most important thermostable polymers is unsaturated polyester resin, due to their physicochemical properties and applications in many industries, such as the construction, textile, automotive and aeronautical, among others. Such resins have been used for elaboration of synthetic fibers, which are versatile and economically accessible, as well as used in composite materials.

Environmental pollution not only occurs when polyesters are discarded, but also during their production and final use. The pollutants are released into the air (gaseous oxides of carbon, sulfur, nitrogen, as well as particulate solids, and even phosphorus), into the water (such as salts and petroleum derivatives), and into the soil (chemicals or toxic salts for animals or plants). Because, they are produced from petroleum, and their extraction is environmentally expensive. But, the greatest concern arises, after they have fulfilled their useful life. It is important to mention, that the world production of polyesters was more than 76 million tons in 2016.

In the manufacturing of polyester products some harmful chemicals are used, for example carcinogens; which are emitted into the air and untreated water, when they are not controlled. It is estimated that most of the plastics thrown into the ocean are polyesters, which come from the textile industry. About 70 billion of oil barrels are necessaries for to meet the demand of them, mainly those polyester for producing synthetic fibers [2]. The largest consumers of polyesters are the Asian countries, where a lot of contaminants have been identified.

As it is known, the chemical reactions involved in the production of unsaturated polyester resins are irreversible and involve a considerable energy consumption; nevertheless, changes on the synthesis ratios can adjust chemical compositions. Then, such resins can be used in many applications.

During the polyester elaboration through to condensation process, the reaction water is produced (by-product), which needs to be treated as waste, for to avoid contamination of the flows. Despite of the advances on the chemical recycling, the waste disposal of such wastes, continue being a contamination problem.

Environmental pollution caused by the textile industry, includes the elaboration of polyester-based textiles, which consists of many operations. In the polyester dyeing processes are analyzed the ecological cost, mainly those related to water consumption, energy expenditure (electrical and thermal), carbon dioxide generation and the effluent toxicity. Wastewater are a serious problem, because they contain several chemical substances (mainly dyestuffs and surfactants), that are harmful to the flora and fauna. In a study, concerning to dyeing polyester, the results show that polyester requires 40 L of water per each kg of dyed material, and more consumption of electrical energy compared to the dyeing of polyamide and cotton. Moreover, thermal energy required for polyester processing is lower compared to other materials [5].

In the industry, there are a lot of non-crosslinked unsaturated polyester resins, whose properties decrease when they exceed the storage time, limiting their use in several applications. The resins can be cross-linked by thermal or chemical catalysis methods. The actions to avoid the cross-linking process includes its reformulation. However, this action compromises its final performance. Moreover, the cost for its reactivation is high, so it is recommendable kept in confinement, and then applying an incineration process [6, 7].

2 Methods for Recycling or Reusing Polyester Resin

As it is known, the most current recycling processes are identified in three groups: mechanical, chemical, and thermal recycling. In the case of mechanical recycling, this in a first stage, follows the shredding and grinding processes; then a method for separation of each component, and finally its application depends on the final use. Such recycling process has been used for to obtain thermostable polymers, which are used as fillers in different materials.

In the thermal recycling, chemical structure and composition are taking into account. Polyester has a disordered macroscopic network, but in some case, this can be partially crystalline. The disordered state is related to the vitreous performance. Moreover, in a molten state, it can have liquid-viscous characteristics. In the case of the recycling of polyester resin, the thermal process requires of temperatures ranging from 300 to 1000 °C. After applying such heat, thermal energy is produced. The processes involved in such treatment involves pyrolysis, gasification or combustion.

Chemical recycling, unlike the mechanical one, is considered like a process that improves the properties of waste materials [3]. Polyester resin recycling are related to degradation processes; which are made with chemicals (organic or inorganic solvents), or aggressive mixtures. The processes involve depolymerization, removal of monomers, and decomposition of the polymer chains [8, 9].

Thermoset polyester can be recovered by subcritical hydrolysis using sodium hydroxide (NaOH) and potassium hydroxide (KOH) as catalyst. A high performance styrene-fumaric acid (SFC) copolymer can be recovered by this process; which has advantages over the thermolysis method; because glass fibers and the inorganic materials are recovered after pyrolysis, however the thermoset resin is not recovered. In this chemical treatment, styrene chains are produced during the cross-linking with polyester molecules. If all ester bonds trapped by styrene chains are hydrolyzed, styrene-fumaric acid copolymer (SFC) is obtained; whose molecular structure is similar to that of styrene maleic acid copolymer, which is a functional polymer. The conditions of the recycling process, include to use 0.38 mol/L of KOH, and two hours of reaction at 230 °C; or use 0.72 mol/L and one hour of reaction at 230 °C. Such conditions generate a resin conversion rate of 82% using NaOH, and 92% by KOH. These values are due to the significant reactivity of subcritical water, which is promoted by its higher ion and low dielectric constant, equivalent to those of organic solvents. The SFC copolymer could be used as raw compound or additive for manufacturing novel resins and functional polymers [6].

Other recycling method consist in the mixture of waste polyester-based materials and phenolic resins, in order to obtain high-performance flame-retardant panels, which are used in the construction and electrical industries. In such composites, particle size of the designer polyester materials must be reduced, for their use as reinforcements. Then, they are mixed with phenolic resins (that act as binders). The resulting composites have low density, high fire resistance, high electrical and thermal isolation; moreover, smoke toxicity can be obtained in case of ignition [10].

Recycling of PET bottles by using natural oleic acid and diethylene glycol is carried out to obtain unsaturated polyester. Oleic acid is a natural fatty acid from different plants, and used in food, cosmetic and pharmaceutical industries, as well as in petroleum applications as additives for pour point improvers. The results show that the physical and curing properties of the unsaturated resins are competitive with the commercial polyester, besides that it is easier to apply, due to its longer gel time. Moreover, they had better thermal stability, and show the conversion of semi-crystalline to amorphous phases. Moreover, its hardness had an improvement of 14%, due to the cross-linking of its polymer chains. Thus, this method is viable, cheaper and friendlier to the environment [11].

Many studies on the depolymerization of polyethylene terephthalate (PET) bottles have been carried out, to reduce their impact on the environment. Chemical recycling of PET have been developed through the processes: (a) alcoholysis, which consist in the transesterification of alcohol with PET at 180–80 °C using an organometallic as catalyst; (b) hydrolysis, where PET is hydrolyzed in water, in an acidic or alkaline medium, at high pressure and temperature, for to obtain scissions of polymer chains; (c) aminolysis, where primary amines reacts with PET from 20 to 100 °C; and zinc acetate, lead or potassium sulfate are used as catalysts; (d) ammonolysis, where ammonia reacts with PET from 70 to 180 °C, at 2 MN/m^2 pressure, and zinc acetate is used as a catalyst; (e) Glycolysis, through PET transesterification reaction from 180 to 240 °C at constant pressure. In the process, the mixture of reagents produces excess of glycolic materials, which can be used for production of new products; which include unsaturated polyester resins, polyurethanes, and epoxy resins, among others; besides is economically viable. Glycolysis is not applicable only to waste unsaturated polyester resin, this can support some technical modifications and be used for different resins. Sometimes, the treated waste may contain calcium carbonate particles or fibers [12].

The chemical process is carried out by a mixture of glycols (ethylene, propylene, diethylene, dipropylene and neopentyl). According to the boiling points of the glycols, thermal process has been carried out from 180 to 250 °C. However, other types of glycols have been used, but the final cost of the product increase, for example (1,3-butanediol, types of bisphenol A). Some catalysts as sodium methylate, sodium ethylate, sodium hydroxide and methanesulfonic acid have been studied.

Thermo-hardened polyester is crushed and then treated with glycols for its easy degradation. Different tools have been used for crushing process, including hammer mills, chain impact, cutting and rollers. Moreover, particle sizes ranging from 200 to 300 μm produce the highest milling efficiency.

At the glycolysis temperature, glass fibers can be recovered from the unsaturated polyester resins. Such recovered fibers can be used as raw material to obtain recycled polyester resins, which have good mechanical properties, high flexibility when they are used as molding material.

3 Gamma Radiation

The gamma radiation is electromagnetic energy, which has been used for decades for modifications of the physicochemical properties of different materials. The ionizing radiation as gamma rays are, has not neither mass nor charge. This is produced from isotope sources, Cesium 137 (^{137}Cs) or Cobalt 60 (^{60}Co) with 0.66 and 1.33 MeV of energy, respectively. Currently, application of the gamma rays is carried out in an inert atmosphere (air or N_2), at the room temperature, covering different irradiation dose and using specific dose rates. The gamma irradiator equipments are produced for limited industries around the world. For example, the 651 PT Gamma beam irradiator which is produced by the Atomic Energy of Canada Ltd, located at Chalk River, Ontario, Canada.

The physicochemical changes produced in polymers by gamma radiation happen through three main processes: (a) scission, or (b) cross-linking of polymer chains, and (c) grafting. Each process is produced according to the physicochemical properties of the polymers and the applied dose rate. In the first stage, the ionizing is procedure on the polymer, through the production of free radicals, that are highly reactive species. Then, such radicals react with neighboring atoms, for to produce either scission or cross-linking of polymer chains. For a total polymerization is necessary to analyze the physicochemical properties of the polymers, including the chemical structure and its molecular weight [13, 14].

The physical and chemical changes provoke by gamma radiation in polymers depend of well-defined parameters, which include: (a) irradiation dose, is to say the absorbed energy per unit mass; and the (b) experimental irradiation conditions, that include the air or nitrogen atmosphere. Several advantages are observed by the gamma radiation process; for example, the chemical reactions can initiate at any temperature and pressure; moreover, this can be applied to different matter states (solid, liquid or gas). Moreover, in such reactions not use of catalysts is carried out. In the case of polymers, the scission and cross-linking processes can modify several physicochemical properties, for example: molecular weight, density, crystallinity degree, thermal behavior, and mechanical resistance [9, 15].

Modifications of the physicochemical properties of materials can carried out by different processes, including those related with temperature, chemicals, or with high energy (ionizing radiation). It is important to mention the advantages of the gamma radiation over thermal and chemical processes: (I) this produces high rates of radical formation, than those produced by current initiators; (II) the reaction begins without the use of catalysts or additives; because this does not require any activation energy; (III) the polymerization can begins at room temperature; (IV) at the initial reaction there is not a significantly increase of temperature. Because, the overall activation energy decrease and do not produce thermal explosions, at difference of those chemical processes; (V) at different stages the temperature and reaction time can be manipulate; (VI) the termination reaction can be controlled, which allows to analyze each stage of the polymerization. The termination stage has several characteristics, for example, if the polymer has high viscosity then high activation energy is produced, thus the overall activation energy decreases (measured in kJ/mol) [14, 16–18].

Moreover, (VII) ionizing radiation produce a complete and fast polymerization; in some of them the catalyst does not have enough energy for a complete polymerization; (VIII) complete polymerization allows better solvent resistance and high shape stability of the polymer dimensions. It is very important to check out the gel and the vitreous states through the process, because in many cases, they happen at the same time; (IX) there is no emission of volatile compounds to the environment [19, 20].

4 Effects of Gamma Radiation in Polyester Resin

At the beginning, the polyester resin is usually in the liquid state. Then, once the reaction temperature is reached, a change of state is produced, from liquid to gel. Then, the polymers chains are bonding between them, which is denominated as a cross-linking process. Chemical reactions are obtained by using some initiators and promoters, whose weight concentration depending on the resin content. Methyl ethyl ketone peroxide (MEKP), amine or benzoyl peroxide (BZP) have been used as initiators, and cobalt naphthenate or dimethyl-para-toluidine (DMPT) as promoters [21, 22].

There are several studies concerning to partial polymerization of polyester resin, is to say, the catalyst is insufficient for such activity. Fortunately, gamma radiation can finish the polymerization process. According to the mentioned advantages, ionizing radiation does not require any activation energy for its initiation; and the final reaction can be controlled. Nevertheless, total polymerization depends of the gamma radiation dose and the type of polymer [17].

During polymerization process, there are reactions between the double-bonds and the free radicals formed on the chains; such process release strain energy. The chain stress is a fundamental parameter, because chemical bond scissions can increase and the recovery radicals decrease, according to the chain stress values. Moreover, there are connection among the parameters: chain length, strain and bond rupture. For example, short chain lengths show high strain energy and fast breaking.

In the first stage of the irradiation process on the polymers, scission of the polymer chains is obtained. Then, radicals are formed in the molecular end groups. Nevertheless, different processes can happen. For example, if there are restrictions of the surrounding chains over the chains with the radicals, then these may rebound with the original chain, is to say, the recovery of the chain is obtained. Such phenomenon is denominated the "cage effects". It is well known, that at low temperature the recovery happen. Then, there is a proportional effect between the total number of the scissions and the number of radicals. But, at high temperature, the "cage effects" happen, is to say, the initial "stress" of the surrounding chains generate "strain" energy on the chains. It is known, that ionizing radiation can help to chains to escape from the restrictions of the surrounding chains. Another phenomenon is called "cage breaks" through which scissions of polymer chains happen first for short chains [23].

Releasing of residual styrene of unsaturated polyester resins generate environmental pollution, in this case those related to the odor. Whereby, gamma radiation is showed as an alternative process for resolve such problem. In the case of irradiated resins, at the first stage free radicals are generate after scissions of polymers chains; then either radical or cationic mechanism can occur. Moreover, the number of chain scissions as well as the release of the local "strain" energy released can improve the mechanical properties [17].

As it is known, the absorbed energy after irradiation process depends mainly of the dose and the molecular weight of the polymers. Nevertheless, sometimes is possible to control the polymerization in terms of the polymer thickness, because at the initial reaction the gamma rays act in homogeneous way throughout the polymer. Moreover, it can happen at low temperature without the use of catalysts or additives. Thus, total polymerization can be controlled in terms of the dose rate [16, 24, 25].

In non-irradiated resins, the polymerization generates rough surfaces with aligned structures, showed by the arrows in Fig. 1a. While, for gamma irradiated resins, the surface is more homogeneous and the aligned structures are more narrow, see the arrows in Fig. 1b [25].

According to the irradiation dose, at 3 kGy, polyester resin is in gel state. At this stage, more cross-linking through to their C=C bonds are produced, and the polyester resin becomes gradually in a viscoelastic gel. But for higher doses, high viscosity is obtained and the gel becomes insoluble, as it is shown for the lines in Fig. 2a. The remaining liquid can be separated from the gel, after the microgels are obtained. They are constituted by spherical particles, as it is shown inside the circles in Fig. 2b, which are produced between polyester insaturations and some styrene molecules [18, 25].

At the total conversion, both the gel fraction and the styrene monomer are present. Is to say, below the reaction temperature different materials are obtained, corresponding to the glass-rubber transition. But, above such temperature, a 3-D network is obtained by interconnection between vinyl monomers and microgels. Thus, polyester resin changes from gel to a hard solid [26].

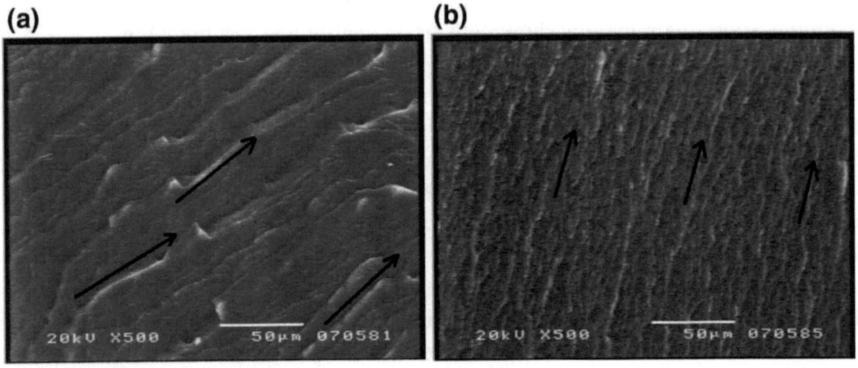

Fig. 1 SEM images of polymerized resin: by using catalyst (**a**), and by gamma rays (**b**)

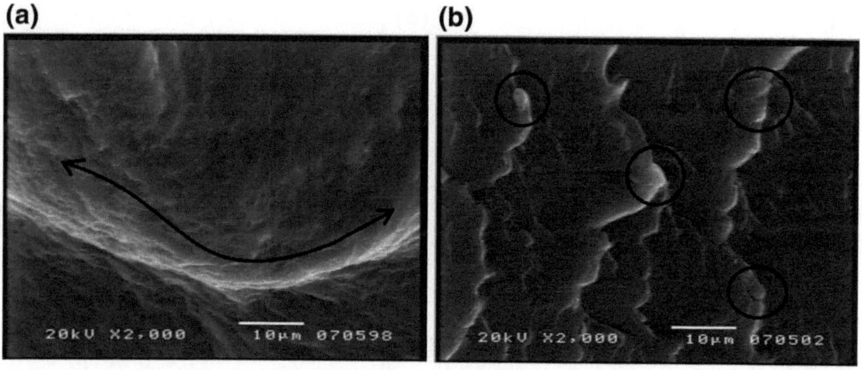

Fig. 2 SEM images of irradiated polyester resin: at 5 kGy (**a**), and 10 kGy (**b**)

The effects of the gamma radiation on the thermal properties of polyester–styrene resin were studied by Thermogravimetric Analysis (TGA). In particular, the decomposition temperature was determined for irradiation doses ranging from 10 to 320 kGy. The results show that 20 kGy are enough for the solidification of polyester–styrene resin. Moreover, the thermal degradation reaction for a weight loss of 10% (T10), was obtained around 330 °C, while at 370 °C weight loss was 30%. Thus, after hardening of polyester resin (at 20 kGy), no significant changes were obtained for the decomposition temperature. In the same study, thermomechanical analyzer (TMA) was utilized to locate glass transition temperature (T_g) of polyester–styrene resins after irradiating; the measurements were obtained with a constant force. The results show, T_g at 67 °C for an irradiation dose of 20 kGy, and a slight increase of 11 °C, when applying 320 kGy. Such increase, is due to the cross-links of polymer chains, which reduces the segmental mobility of the chains [27].

According to the literature, glass transition temperature, T_g, increases when irradiation dose increases. For example, in the case of non-irradiated polyester resins, The T_g was located at 260 °C with a weight loss of 5%, but for irradiated polyester, T_g increases 30 °C. In the case of the decomposition temperature, the values for both, non-irradiated and irradiated polyester resin, are very similar. Nevertheless, the values are higher in presence of nitrogen than those obtained in oxygen [28–30].

After irradiating, stiffness of polyester resin gradually increases, as it is shown in Fig. 3. For non-irradiated resin, smooth surface is obtained, but for an irradiation dose of 150 kGy, the surface changes, now is more rough with some cracks, which were obtained after scissions of polymer chains [31].

Mechanical properties or gamma irradiated polyester-styrene resin were evaluated. The results show that the stress at break increases when irradiation dose increases. At 20 kGy the value is 6.5 MPa, which increase 38% for an irradiation dose of 320 kGy. Polymer chains build cross-links between the chains, thus the density increases. Such density increases with increasing the irradiation dose.

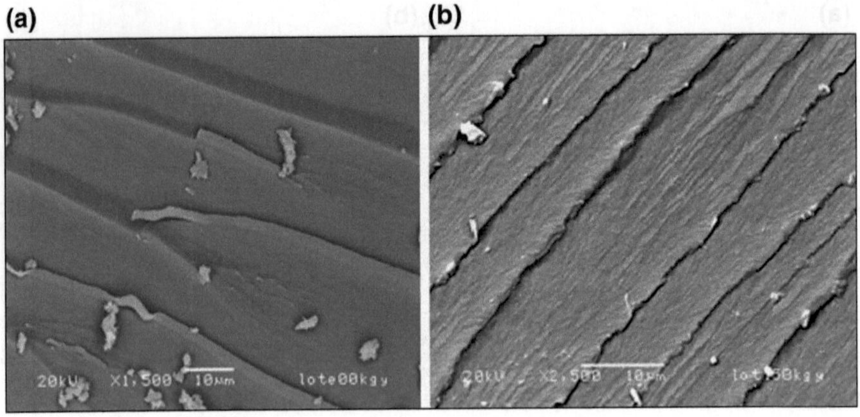

Fig. 3 SEM images of non-irradiated and irradiated polyester resin

Nevertheless, the hardness of polyester-styrene resin is not affected, at 20 kGy the value is 89%, which decrease 88% at 320 kGy [27].

Dielectric and thermal analysis of unsaturated polyester resin were obtained by measurements by stepwise changing of frequency and temperature. The results show that the glass transition temperature, T_g, increases when the frequency or the cross-link density increase. The apparent activation energy is obtained at about 8 kJ/kg, while the glass transition temperature increase at about 13 kJ/kg. The lowest transition corresponds to the exothermic peak, which is highly asymmetric. For high reaction time, the dielectric permittivity decreases [32].

The uncured polyester resins show a typical asymmetric shape with no indication of heterogeneity. But, their phase-states are continuously changing during their curing. They show two transitions with relative intensities changing as curing proceeds. At a very early stage of the curing process the gel state is obtained. In the case of irradiated cured resins at least two transitions with inhomogeneous characteristics are obtained. Respect to the dielectric studies; the shape of the dielectric spectrum peaks indicated a heterogeneous nature of the transition, which increase when the degree of curing increase. Moreover, the permittivity decreases due to the cross-linking of polymer chains, and to the fact that the dielectric transition is swept through the reaction temperature by the curing reaction. In the uncured and partially cured resins, at low temperatures, the unrelaxed reaction time limit of the permittivity is lower than that for long reaction time limit [32].

5 Gamma Radiation as Modifier of Composites Elaborated with Polyester Resins

Composites elaborated with polyester resin and different polymers, acting as reinforcements or fillers, have been elaborated. They show high values on the mechanical properties; chemical and corrosion resistance; thermal stability; good electrical conductivity; easy processability, durability, as well as low density and cost. Reinforced polyester resins have been used in household products, biomedical applications, building materials, automotive and aircraft industry, spacesuits and sports goods.

In a study, a polyester base composite was elaborated adding raw jute fabrics (a lingo-cellulosic fiber crop), by the heat-press molding technique. It was composed of three layers of jute fabrics and polyester resin mixed with cross-linking agents. Such composites were irradiated from 2 to 14 kGy, for the complete cross-linking of the polyester resin. The results show high degree of cross-linking and morphological changes on the surface of the composites as well as highest improvements on the physical and mechanical properties when they were irradiated at 10 kGy. In particular, the tensile strength increase from 60 to 90 MPa, and the flexural strength from 92 to 154 MPa. As it is known, the higher stiffness of the raw jute fabric is due to the higher lignin content, this imparts higher tensile strength and tensile modulus to the composite [33].

As higher the values of tensile and flexural strength, more stiffness and low flexibility is obtained. In the case of irradiated composites, the tensile modulus increase from 1.8 to 2.9 GPa, the flexural modulus from 3.1 to 5 GPa, but they showed low flexibility; such results may be attributed as the increased in cross-linking extend in the matrix by irradiation, which made the composites more ductile. Moreover, the surfaces of the composite (analyzed by AFM), showed smoother morphology for non-irradiated composites, instead of those for irradiate ones. Radiation induces highly dense cross-linking in the matrix, which improve the compactness and increase the roughness of the surfaces [33].

Composites elaborated with unsaturated polyester resin and marble show changes on their morphologies, according to irradiation gamma dose. In Fig. 4, SEM images of non-irradiated composites show a smooth and homogeneous surface with some marble particles covered by polyester resin, but after irradiating with gamma rays, morphological changes are obtained, the polyester resin shrinks and the marble particles agglomerate. More detachable particles are obtained and the presence of small cavities less than 10 μm are observed.

Composites with polyester resin and jute treated fabric, were elaborated by hand lay-up and heat press molding techniques. After, they were irradiated from 2 to 9 kGy. The results show, the highest improvements at 5 kGy of irradiation dose. The tensile strength improves 31%, while the bending strength 26%, and impact strength 12%, such values respect to the values for non-irradiated composites. Such increase may be due to the intercross-linking between the neighboring cellulose molecules, which resulted in the strength of natural fiber. But at higher irradiation dose, the main chain may be broken down and polymer may degrade into fragments and, as a result, mechanical properties will decrease. Nevertheless, low water uptake as well as higher moisture resistance of jute polyester composites are ideal for roofing and wall cladding in industrial building. Moreover, they can primarily be used for low-cost housing and automotive interior component applications [34].

Composites elaborated with polyester resin (50%) and PET fibers were elaborated by hand lay-up and compression molding process. After they were irradiated with

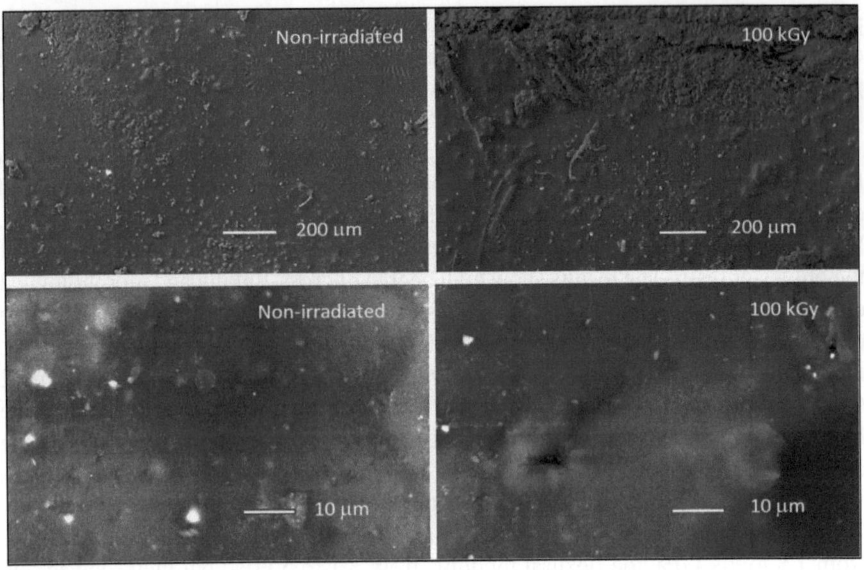

Fig. 4 SEM images of non-irradiated and irradiated polyester resin/marble composites

gamma rays at 1, 2.5, 5 and 10 kGy at a dose rate of 3 kGy/h. The results show improvements on the mechanical properties. The highest mechanical values were obtained at 2.5 kGy. The tensile strength improves 4.53%, this increased up to certain gamma irradiation dose and then decreased. The tensile modulus improves 31.32%, bending strength 22.87%, bending modulus 24.96%. In the case of impact strength, the highest values were obtained at 10 kGy, which means an improvement of 28.21%. Gamma irradiation may also remove moisture from the composite, which in turn contributed to better fiber matrix adhesion. Thus gamma irradiation has influence on the interfacial bond strength, by producing active sites inside the matrix and the fibers. This, may be the reason behind the increased mechanical properties of the composites [35].

6 Conclusions

In this chapter was discussed the importance of the gamma radiation as a novel tool for recycling of polyester resins, and their applications for the improvement of the physicochemical properties of composite materials. The results show that such ionizing energy can be viable as solution of the environmental problems caused by waste polyesters, mainly for those related to the manufacturing of environmental friendly materials. Nevertheless, more extensive investigations are required in the future, for to consolidate to the gamma radiation as an innovative and clean technology.

References

1. Tawfik ME, Eskander SB (2006) Polymer concrete from marble wastes and recycled poly(ethylene terephthalate). J Elastom Plast 38:65–79. https://doi.org/10.1177/0095244306055569
2. Hoornweg D, Bhada-TaTa P (2012) What a waste. A global review of solid waste management. Urban development series. Knowledge papers. World Bank, Washington, DC, USA. ISBN 978-1-4648-1329-0
3. Oliveux G, Bailleul J-L (2013) Recycling of glass fibre reinforced composites using subcritical hydrolysis: reaction mechanisms and kinetics, influence of the chemical structure of the resin. Polym Degrad Stab 98:785–800. https://doi.org/10.1016/j.polymdegradstab.2012.12.010
4. Yang Y, Boom R, Irion B, Heerden D, Kuiper P, Wit H (2012) Recycling of composite materials. Chem Eng Process 51:53–68. https://doi.org/10.1016/j.cep.2011.09.007
5. Rosa JM, Garcia VSG, Boiani NF, Melo CG, Pereira MC, Borrely SI (2019) Toxicity and environmental impacts approached in the dyeing of polyamide, polyester and cotton knits. J Environ Chem Eng 7:102973. https://doi.org/10.1016/j.jece.2019.102973
6. Nakagawa T, Goto M (2015) Recycling thermosetting polyester resin into functional polymer using subcritical water. Polym Degrad Stab 115:16–23. https://doi.org/10.1016/j.polymdegradstab.2015.02.005
7. Lu T, Solis-Ramos E (2016) Synergistic environmental degradation of glass reinforced polymer composites. Polym Degrad Stab 131:1–8. https://doi.org/10.1016/j.polymdegradstab.2016.06.025
8. Goodship V (2010) Management, recycling and reuse of waste composites. Woodhead Publishing Ltd., Cambridge, UK and CRC Press, Boca Raton, FL, USA. ISBN 978-184-56-9462-3
9. Blazsó M (2010) Pyrolysis for recycling waste composites. In: Goodship V (ed) Management, recycling and reuse of waste composites. Woodhead Publishing Ltd., Cambridge, UK and CRC Press, Boca Raton, FL, USA, pp 102–121. ISBN 978-184-56-9462-3
10. Correa M, Laza JM, Vilas JL, Bilbao E, Rodríguez M, León LM (2010) Reutilization of thermostable polyester wastes by means of agglomeration with phenolic resins. Waste Manage 30:2305–2311. https://doi.org/10.1016/j.wasman.2010.05.007
11. Naguib HM, Zhang XH (2018) Advanced recycled polyester based on PET and oleic acid. Polym Testing 69:450–455. https://doi.org/10.1016/j.polymertesting.2018.05.049
12. Raheem AB, Noor ZZ, Hassan A, Hamid MKA, Samsudin SA, Sabeen AH (2019) Current developments in chemical recycling of post-consumer polyethylene terephthalate wastes for new materials production: a review. J Clean Prod 225:1052–1064. https://doi.org/10.1016/j.jclepro.2019.04.019
13. Martínez-Barrera G, Menchaca Campos C, Ureña-Nuñez F (2012) Gamma radiation as a novel technology for development of new generation concrete. In: Adrovic F (ed) Gamma radiation. InTech, Rijeka, Croatia, pp 91–114. ISBN 978-953-51-0316-5
14. Martínez-Barrera G, Brostow W (2009) Fiber-reinforced polymer concrete: property improvement by gamma irradiation. In: Barrera-Díaz C, Martínez-Barrera G (eds) Gamma radiation effects on polymeric materials and its applications. Research Signpost, Kerala, India, pp 27–44. ISBN 978-81-308-0293-0
15. Martínez-Barrera G, Menchaca-Campos C, Barrera-Díaz CE, Avila-Cordoba LI (2013) Recent developments in polymer recycling. In: Bikit I (ed) Gamma rays: technology, applications and health implications. Nova Science Publishers Inc., Hauppauge, NY, USA, pp 237–256. ISBN 978-1-62257-697-5
16. Cruz-Zaragoza E, Martínez-Barrera G (2009) Ionizing radiation effects on the matter and its applications in research and industry. In: Barrera-Díaz C, Martínez-Barrera G (eds) Gamma radiation effects on polymeric materials and its applications. Research Signpost, Kerala, India, pp 1–14. ISBN 978-81-308-0293-0
17. Clough RL (2001) High-energy radiation and polymers: a review of commercial processes and emerging applications. Nucl Instrum Methods Phys Res B 185:8–33. https://doi.org/10.1016/S0168-583X(01)00966-1

18. Jurkin T, Pucic I (2006) Post-irradiation crosslinking of partially cured unsaturated polyester resin. Radiat Phys Chem 75:1060–1068. https://doi.org/10.1016/j.radphyschem.2006.04.001
19. Chapiro A (2002) Polymer irradiation: past–present and future. Radiat Phys Chem 63:207–209. https://doi.org/10.1016/S0969-806X(01)00621-1
20. Brostow W, Glass NM (2003) Cure progress in epoxy systems: dependence on temperature and time. Mater Res Innov 7:125–132. https://doi.org/10.1007/s10019-002-0222-2
21. Guedes RM, Tavares CML, Ferreira AJM (2004) Experimental and theoretical study of the creep behavior of GFRP-reinforced polymer concrete. Compos Sci Technol 64:1251–1259. https://doi.org/10.1016/j.compscitech.2003.10.004
22. Ahn N (2003) Effects of diacrylate monomers on the bond strength of polymer concrete to wet substrates. J Appl Polym Sci 90:991–1000. https://doi.org/10.1002/app.12723
23. Nishiura T, Nishijima S, Okada T (1999) Creep behavior of epoxy resin during irradiation at cryogenic temperature. Radiat Phys Chem 56:605–609. https://doi.org/10.1016/S0969-806X(99)00294-7
24. Martínez-Barrera G, Ureña-Nuñez F, Gencel O, Brostow W (2011) Mechanical properties of polypropylene-fiber reinforced concrete after gamma irradiation. Compos Part A Appl Sci Manuf 42:567–571. https://doi.org/10.1016/j.compositesa.2011.01.016
25. Martínez-Barrera G, Vigueras-Santiago E, Gencel O, Hagg Lobland HE (2011) Polymer concretes: a description and methods for modification and improvement. J Mater Ed 33:37–52
26. Dispenza C, Alessi S, Spadaro G (2008) Carbon fiber composites cured by γ-radiation induced polymerization of an epoxy resin matrix. Adv Polym Technol 27:163–171. https://doi.org/10.1002/adv.20127
27. Ajji Z (2005) Preparation of polyester/gypsum/composite using gamma radiation, and its radiation stability. Radiat Phys Chem 73:183–187. https://doi.org/10.1016/j.radphyschem.2004.08.004
28. Ismail MR, Ali MA, El-Milligy AA, Afifi MS (1998) Physico-chemical studies of gamma-irradiated polyester-impregnated cement mortar composite. J Radioanal Nucl Chem 238:111–117. https://doi.org/10.1007/BF02385364
29. Czayka M, Fisch M, Uribe RM, Vargas-Aburto C (2007) Radiation-thickening of iso-polyester resin. Radiat Phys Chem 76:1058–1068. https://doi.org/10.1016/j.radphyschem.2006.10.007
30. Pucic I, Ranogajec F (2003) Phase separation during radiation crosslinking of unsaturated polyester resin. Radiat Phys Chem 67:415–419. https://doi.org/10.1016/S0969-806X(03)00077-X
31. Martínez-Barrera G, Espinosa-Pesqueira ME, Brostow W (2007) Concrete + polyester + $CaCO_3$: mechanics and morphology after gamma irradiation. e-Polymers 7:083. https://doi.org/10.1515/epoly.2007.7.1.956
32. Jelcic Z, Hedvig P, Ranogajec F, Dvornik I (1982) Dielectric and thermal analysis of radiation curing of unsaturated polyester resins. Radiat Phys Chem 20:309–314. https://doi.org/10.1016/0146-5724(82)90120-0
33. Hoque MA, Bhuiya MK, Saiduzzaman M, Islam MA (2005) Preparation of raw jute fabric reinforced and low lignin content modified jute fabric reinforced polyester composites-effects of gamma radiation on their properties. In: International conference on materials, electronics & information engineering. www.ru.ac.bd/icmeie2015/proceedings/
34. Ayma A (2017) Effect of gamma radiation on the properties of jute reinforced polyester matrix composites. J Text Sci Eng 6:294–296. https://doi.org/10.4172/2165-8064.1000294
35. Das SC, Paul D, Islam JMM, Khan MA (2016) Effect of gamma radiation on the mechanical properties of PET felt reinforced polyester composites. In: International conference on mechanical, industrial and energy engineering, Bangladesh

Studies on Recycled Polyester

Yijing Qin, Muchao Qu, Joachim Kaschta, Vincent Allen and Dirk W. Schubert

Abstract In this chapter, poly(ethylene terephthalate) (PET) melt-spun fibres from five different PET materials were fabricated by a capillary rheometer, which was equipped with an air aspirator for aerodynamic stretching. Two recycled PET of coloured ***rPET-A*** and clear ***rPET-B*** from waste PET bottle flakes were used, and three virgin PET with different grades of the bottle-grade ***vPET-2***, fibre-grade ***vPET-1*** and fibre-grade ***vPET-3*** were chosen as reference benchmark. During the extrusion process, two processing temperatures of 270 and 280 °C in the capillary rheometer were applied, combined with six different take-up pressures from 0.5 to 3.0 bar for fibre stretching during the spinning process. An empirical equation was given, to predict the water content of PET materials under the specific drying temperature and drying time. Afterwards, the characteristics of the raw materials and drawn PET fibres were discussed comprehensively from the perspective of thermal behaviour, molar mass characteristic, rheological properties, surface morphology, diameter, tenacity and elongation at break. ***rPET-B*** and ***vPET-1*** materials yielded optimal spinnability and demonstrated to be the most suitable materials to fabricate fibres with a desirable tenacity, an outstanding elongation at break and an excellent fibre fineness. Crystallinity and orientation of fibres from these two materials were explored to give a potential explanation for the higher tenacity. Furthermore, performance-reliability plots were applied, and a semi-empirical equation was suggested for the first time relating the tenacity and elongation at break of fibres. This study proves the possibility to produce PET fibres from bottle-grade recycled co-polymer PET material, which have similar properties as fibres produced from fibre-grade virgin homo-polymer PET. All the specific information of the five investigated PET materials, especially ***rPET-B*** and ***vPET-1***, can provide guidance for melt spinning process in the future.

Y. Qin · M. Qu · J. Kaschta · V. Allen · D. W. Schubert (✉)
Institute of Polymer Materials, Friedrich-Alexander-University Erlangen-Nuremberg, Martensstr. 7, 91058 Erlangen, Germany
e-mail: dirk.schubert@fau.de

Y. Qin
e-mail: yijingqin1991@gmail.com

Y. Qin · M. Qu · D. W. Schubert
KeyLab Advanced Fiber Technology, Bavarian Polymer Institute, Dr. Mack-Strasse 77, 90762 Fürth, Germany

S. S. Muthu (ed.), *Recycled Polyester*, Textile Science and Clothing Technology,
https://doi.org/10.1007/978-981-32-9559-9_3

Keywords Recycled PET · Virgin PET · Melt spinning · Thermal behaviour · Viscosity · Surface morphology · Mechanical property · Semi-empirical equation

1 Introduction

In the 1970s, there was an unprecedented usage of poly(ethylene terephthalate) (PET) as a packaging material, when one highly oriented PET container was produced from high-molar-mass PET by the Du-Pont company [1]. Since then, many factors, including good mechanical properties, chemical resistance, clarity, thermal stability, light-weight, easy disposability and recyclability, have been realized and have motivated PET to be a preferable material in the packaging industry [2–9]. PET is extensively used for the manufacturing of liquid containers, especially PET bottles used for mineral water, bubble water, carbonated drinks, as well as other beverages. As well documented, the proportion of PET bottles in the global PET resin requirement is 83–84% [10]. The growth of PET consumption together with the slow natural decomposition rate of PET has resulted in a serious environmental pollution problem. There is also a low utilization efficiency of post-consumer PET bottles [11, 12]. Due to growing attention to environmental protection and the better utilization of the resource, PET recycling has played a significant role in the global supply chain of PET bottles [13, 14]. The recycling process of PET bottles would provide the potential to reduce fossil fuel and energy consumption while also reducing greenhouse gas emissions [15].

Techniques for recycling PET bottles consist of chemical recycling, physical recycling, and energy recovery (i.e., burning) [8, 16–19]. Chemical recycling is a high-cost method which includes the depolymerization of the post-consumer PET waste into monomer units or randomly into chain fragments. The formation of gas by-products is normally associated with this depolymerization process. The obtained monomers are then purified using the distillation method and dried for generating new polymers. In physical recycling, namely mechanical recycling, PET bottles are separated from other polymer bottles and contaminations, and then are washed before the smashing process. Eventually, PET bottle flakes or re-extruded PET granules from extrusion moulding are obtained. This physical recycling process causes deteriorations of properties to some extent, for instance, the discolouration or molar mass loss resulting from chain scission or hydrolysis with the existence of water, acidic pollutions or other polymers. Therefore, until now only a small amount of recycled PET from post-consumer bottles are employed for the manufacture of original products, while the majority of them are employed to produce fibres for blankets or clothing, without recycling again after use [20].

In the past two decades, a considerable amount of literature has been published on the manufacturing of fibres from recycled PET waste through different processing routes, such as melt spinning [1, 18, 19, 21–26], electrospinning [9, 27–29], or newly-invented centrifugal spinning [30]. As known, the physical properties of fibres are significant criteria to evaluate their potential applications. Focusing on the most prevalent melt spinning, Abbasi et al. [19] investigated the influence of take-up speeds on the physical performance of virgin PET and PET bottles flakes, where an automatic spinning machine was used. However, the authors failed to note the extrusion speed, rendering the results incomparable and unreliable. Different PET blends with the recycled PET and virgin PET grades were studied by Gurudatt et al. [1]. Unfortunately, in this work, many variables were introduced because individual processing temperatures were used for the different recycled/virgin PET ratios during melt spinning. In the study reported by Elamri et al. [25], the thermal and rheological behaviour of recycled/virgin PET blends were characterized; nevertheless, only one measured specimen was given to discuss the mechanical properties, and no diameter value of the investigated fibres was provided. With respect to the virgin/recycled PET blends, Lee et al. [18] also researched the influence of the different percentage of virgin PET on the physical properties, but only one draw ratio was applied.

In the aforementioned literature survey on melt spinning, it should be realized that a systematic paper attempting to explore the physical properties, as well as the spinnability and fibre surface of virgin and recycled PET with various processing parameters was missing. Considering this, a research-based and general investigation was first reported by Qin et al. [12] to assess the potential of producing melt-spun fibres using recycled and virgin PET materials.

In this chapter, a more comprehensive and complementary research work on the basis of the previous study [12] is presented. In Sect. 2, the investigated materials, preparation processes and analysis methods used in this study are listed. The Characterizations of PET Raw Materials are discussed in Sect. 3, where the thermal, molar mass and rheological behaviour of raw materials are examined to deeply explore properties relevant to melt spinning. In Sect. 4, the thermal and molar mass behaviour of melt-spun fibres are investigated as well to make a comparison with those of raw materials. In addition, the surface morphology, diameter distribution, thermal stability, mechanical properties of melt-spun fibres produced from different PET materials are discussed and compared. Performance-reliability Plot for the two most desirable PET materials are shown in Sect. 5 and the Comparison with Reviewed Literature are addressed in Sect. 6.

2 Experimental Part

2.1 Materials Preparation

Two kinds of recycled PET flakes from post-consumer PET bottles in Fig. 1a (copolymers, named ***rPET-A*** and ***rPET-B***) were supplied by TEXPLAST GmbH, Germany. ***rPET-A*** contains different coloured flakes (white flakes ca. 36%, brown 22%, green 21%, blue 18%, and others 3%), and ***rPET-B*** consists of clear flakes and a very small portion of blue flakes (less than 5%). Compared to the information provided by the supplier, the PET recycling process was consistent with physical recycling

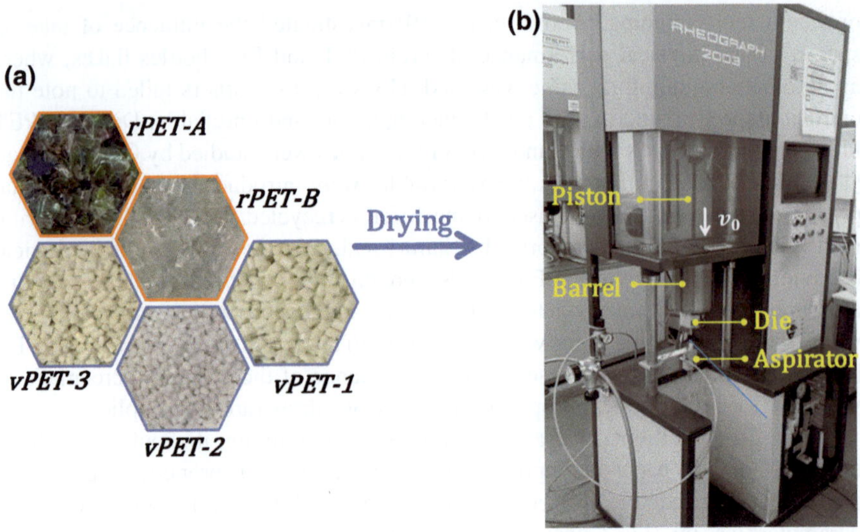

Fig. 1 **a** Appearance of the five investigated PET materials and **b** the fibre spinning apparatus

as described above. Further properties and possible contaminations of the recycled flakes are shown in Table 1. From this chart, some impurities such as polyethylene (PE), polypropylene (PP), polyvinyl chloride (PVC), polyamide (PA), paper, and metals can be found in two recycled PET flakes. In general, the purity of ***rPET-B*** is higher than that of ***rPET-A***, which explains the larger molar mass degradation of ***rPET-A*** after processing into fibres.

Table 1 The properties and contamination information of ***rPET-A*** and ***rPET-B*** flakes

	rPET-A	***rPET-B***
Property		
Apparent density	≥250 kg/m^3	≥250 kg/m^3
Flake size	≤10 mm	≤10 mm
Moisture content	≤1.2%	≤1.0%
Dust (<0.5 mm)	≤0.5%	≤0.5%
Contamination		
Polyolefin (PE/PP)	≤80 ppm	≤30 ppm
PVC	≤50 ppm	≤20 ppm
PA	≤200 ppm	≤50 ppm
Paper/labels	≤100 ppm	≤50 ppm
Metal	≤50 ppm	≤10 ppm
Others	≤200 ppm	≤100 ppm

Rearranged by authors from Ref. [12]

Table 2 The INVISTA's information on virgin PET granules

	IV[a] (dL/g)	Type	Application
vPET-1	0.64	Homo-polymer	Fibre-grade
vPET-2	0.81	Co-polymer	Bottle-grade
vPET-3	0.89	Homo-polymer	Fibre-grade

[a]Obtained by the conversion of solution viscosity into intrinsic viscosity, with an empirical relationship proposed by INVISTA, which is equivalent to ASTM method D-4603
Rearranged by authors from Ref. [12]

One should notice that in this study, the recycled PET flakes for fibre spinning were originally from PET bottles, i.e., bottle-grade PET. Therefore, to provide a good reference benchmark for the recycled PET flakes, three virgin PET granules, composed of two fibre-grades and one bottle-grade, were selected (INVISTA Resins & Fibers GmbH, Germany). The investigated virgin PET granules were deliberately chosen according to the intrinsic viscosity (IV), which acted as the most important factor to evaluate a polymer in fibre melt spinning and nonwovens industry. As reported in the literature [1, 9, 18, 19, 21, 30–32], a varying range of IV from 0.52 to 1.05 dL/g has been used for PET fibre spinning. Thus, as summarized in Table 2, three different IV, incremented from 0.64 to 0.89 dL/g, for the virgin PET granules are determined. The virgin PET granules are shown in Fig. 1a as well.

2.2 *Karl Fischer Titration*

To prevent the PET molar mass from hydrolytic chain scission in the melt spinning process, water content of PET materials must be less than 0.005 wt% (i.e., 50 ppm) [33–36]. The water content after drying was determined using Karl Fischer titration (C30 Coulometric KF, Mettler-Toledo, Germany) with an oven temperature of 200 °C. As presented in Table 3, the as-received recycled PET flakes have nearly ten times higher water content than that of virgin PET granules. The higher absorption of the other contaminants in the recycled PET flakes is identified as the probable major contributing factor for this phenomenon, such as PA (see Table 1). Consequently,

Table 3 The water content of different PET materials before and after drying, as well as the *predicted value* of the virgin PET materials using Schubert Eq. (1)

	Average water content (ppm)				
	rPET-A	***rPET-B***	***vPET-1***	***vPET-2***	***vPET-3***
Without drying	8807.2	7618.5	510.7	769.2	296.1
353.15 K for 68 h	328.1	317.0	14.6	17.5	9.8
Predicted value	–	–	*21.1*	*31.7*	*12.2*
413.15 K for 48 h	36.8	48.1			

when the water content of virgin PET reached a qualified value (<50 ppm) at 80 °C (353.15 K) for 68 h, the recycled PET flakes still contained much more water inside (>300 ppm). To achieve the required water content for the PET flakes, as well as to shorten the drying time, a higher temperature of 140 °C (413.15 K) was needed. To ensure the same thermal history before melt spinning, the drying procedure at 140 °C for 48 h was employed for all five, as-received, materials.

Accordingly, an empirical Eq. (1) was proposed by Schubert to predict the water content of PET materials at different drying temperature with different drying time:

$$H = H_0 \cdot e^{-t/\tau} \tag{1}$$

with

$$\tau = A \cdot e^{B/T} \tag{2}$$

where H_0 is the water content of as-received PET material (in ppm), and H refers to the real water content after drying at temperature T in Kelvin scale with drying time t (in hour); τ is a parameter dependent on the drying temperature; A and B are constant, which are in relation to the properties and geometry of the measured polymer materials. To predict the values of A and B, the data of ***rPET-A*** and ***rPET-B*** obtained from two drying temperatures and the corresponding time were used, and the results are shown in Table 4.

It should be noted that the values of A and B of both recycled PET flakes are close to each other. To further verify the feasibility, the water content of three virgin PET materials dried at 80 °C for 68 h were predicted by combining Eqs. (1) and (2) with the average values listed in Table 4. Comparing with the actual measured values in Table 3, the predicted values are acceptable considering the measurement accuracy when the water content is below 50 ppm. Overall, the empirical equation provides only a rough estimation which helps to design the drying temperature and the drying time to remove the water content of the PET materials. The reliability of this equation needs to be demonstrated by further experiments in detail, which is left for future studies.

Table 4 Calculated A and B, as well as the average values of ***rPET-A*** and ***rPET-B***

	A (h)	B (K)
rPET-A	0.0561	2086.73
rPET-B	0.0787	1979.35
Average value	0.0674	2033.04

2.3 Melt Spinning in Laboratory Scale

Before the melt spinning, all the PET materials were completely dried to remove the moisture. A capillary rheometer (Rheograph 2003, Göttfert, Germany) equipped with an air aspirator was used for PET melt spinning (Fig. 1b). The air aspirator was installed directly 16 cm below the die, on the same centerline. The optimized experimental parameters have been explored in the preliminary work, which are: the die with length L of 10 mm and diameter D of 0.5 mm; extrusion speed of the piston v_0 equal to 0.1 mm/s. Therefore, according to Eq. (3) [37, 38], the apparent shear rate $\dot{\gamma}_a$ of the PET materials in this study can be thus obtained as in $\dot{\gamma}_a = 921.6\ \mathrm{s}^{-1}$.

$$\dot{\gamma}_a = 4Q/\pi R^3 = 8 \cdot v_0 \cdot D_0^2/D^3 \quad (3)$$

where Q refers to the throughput of the materials in the extrusion direction, and R is the radius of the die used during the extrusion process, and D_0 refers to the diameter of the piston, with $D_0 = 12$ mm.

Two processing temperatures of 270 and 280 °C were applied for the melt spinning process in this study. Considering the difference on T_m for both types of polyester (homo- and co-polymer) was about 10 °C (Table 5), hence these two chosen temperatures exactly provide a comparison of both kinds of PET materials at 26 °C above T_m. In order to obtain a complete molten state of the PET flakes or granules, a melting time of 480 s prior to the melt spinning process was given, which was explored as the most optimal parameter in the preliminary work. Once the PET filaments were extruded out of the die, they were directly introduced through the aspirator, where the filament was aerodynamically stretched by the high-speed flowing air. The investigated take-up pressures of the aspirator were given as 0.5, 1.0, 1.5, 2.0, 2.5 and 3.0 bar, respectively.

Therefore, the effective take-up speed (v_F) and draw ratio (DR) can thus be calculated as:

Table 5 Thermal properties of the PET raw materials derived from Fig. 2

Specimens	T_g (°C)	T_{m1} (°C)	T_{m2} (°C)	ΔH_m (J/g)	ΔH_c (J/g)	X_c (%)
rPET-A/Clear	80.5	–	243.4	29.6	6.0	17.4
rPET-A/Blue	80.4	235.8	244.3	29.1	–	21.4
rPET-A/Brown	80.1	234.4	244.8	30.9	–	22.7
rPET-A/Green	81.0	232.3	243.7	27.2	–	20.1
rPET-B/Clear	80.1	–	243.5	28.9	6.4	16.6
rPET-B/Blue	80.3	234.8	244.5	29.5		21.7
vPET-1	82.8	239.2	253.6	33.6	–	24.8
vPET-2	80.0	–	243.6	28.6	6.4	16.4
vPET-3	81.9	–	251.7	28.6	6.9	16.0

Rearranged by authors from Ref. [12]

$$Q = v_0 \cdot \pi \cdot (D_0/2)^2 = v_E \cdot \pi \cdot (D/2)^2 = v_F \cdot \pi \cdot (D_F/2)^2 \tag{4}$$

$$v_F = v_0 \cdot (D_0/D_F)^2 \tag{5}$$

$$DR = v_F/v_E = D^2/D_F^2 \tag{6}$$

where Q is the throughput in the extrusion process, v_0 is the extrusion speed of the piston and D_0 is the diameter of the piston, which are the same as defined in Eq. (3). v_E denotes the speed of extruded filaments when they leave from the die; D_F denotes the diameter of drawn PET fibres which are collected by a perforated plate below the aspirator for further analysis. As calculated, the range of take-up speed v_F in this study is varying from ca. 60 to 3500 m/min, which is different from each PET material at the same take-up pressure due to the different elongational viscosities. The draw ratio DR of all the drawn fibres varies from ca. 20 to 1000 with the take-up pressure increasing from 0.5 to 3.0 bar.

2.4 *Differential Scanning Calorimetry*

The melting behaviour and crystallization properties of all the investigated PET grades, as well as the representative PET fibres, were examined using a differential scanning calorimetry (DSC, TA Q2000, TA Instruments, USA) in a nitrogen atmosphere. For the PET raw materials, the specimens with the weight of ca. 6 mg were heated up to 300 °C with a rate of 10 °C/min. Afterwards, the specimens were kept for 3 min at 300 °C before cooled down to 25 °C with the same rate. The same cycle (heating up and cooling down) was repeated again, aiming at erasing the thermal history of PET material when processing. For the PET fibres, the first heating scan was used to obtain the glass transition temperature T_g, melting temperature T_m, and crystallinity X_c influenced by different processing parameters. The crystallinity X_c was calculated by the following equation [39]:

$$X_c = (\Delta H_m - \Delta H_c)/\Delta H_m^0 \tag{7}$$

where the ΔH_m denotes the enthalpy of fusion for the specimen at the melting point, ΔH_c denotes the enthalpy of recrystallization, and ΔH_m^0 denotes the enthalpy of fusion for a 100% crystalline polymer at the equilibrium melting point, which is reported as 135.8 J/g for PET [40].

2.5 Morphological Study and Diameter Evaluation

To evaluate the fibre surface roughness, the PET fibres generated at 270 °C and 0.5 bar were representatively selected to be performed the scanning electron microscopy observation (SEM, AURIGA, Carl Zeiss Microscopy, Germany). Furthermore, the overall morphology and diameters of the investigated PET fibres were analyzed by another optical microscope (VHX-J250 Keyence, Japan).

2.6 Single Fibre Tensile Test

The mechanical properties in the materials characterizations are of great significance to evaluate the applications of the products. In this work, the tenacity and elongation at break of drawn PET fibres were investigated using a single fibre tensile tester (Vibrodyn 400, Lenzing Instruments, Austria). The continuous melt-spun PET fibres were cut into specimens with a length of about 50 mm for the tensile test. For each group of fibres produced under one combined processing condition (one material, one processing temperature, and one take-up pressure), at least 20–30 specimens were prepared and used to obtain the average value and standard deviation. Before the tensile test, the specimen between the rubberized clamps was carefully straightened with a tiny initial force of 0.2 cN (1 cN = 0.01 N).

2.7 Hot Shrinkage

Due to the low crystallinity of PET material, the orientation of molecular chains is mainly located in the amorphous zone. Therefore, the hot shrinkage measurement was considered as the most suitable method to quantify the orientation. The hot shrinkage ratio (SR) after thermal treatment, defined as the variation of percentage in the length of the fibres can be obtained as presented in Eq. (8):

$$SR = (L_0 - L_1)/L_0 \times 100\% \qquad (8)$$

where L_0 and L_1 are the lengths of initial PET fibre and the corresponding value after thermal treatment, respectively. The loose PET fibres were treated in an oven with a temperature of 90 °C for 60 min. The temperature of 90 °C was employed, which is located in the range from the glass transition temperature T_g to the onset recrystallization temperature $T_{c,onset}$ of all the studied PET fibres (see the DSC data in Table 9). Thus, the shrinking effect due to recrystallization on the fibre length can be avoided. For PET fibres produced under each processing condition, at least ten specimens were investigated by taking pictures before and after the thermal treatment. The initial length and the length after thermal treatment were analyzed using the software of JMicrovision. Note: all the investigated PET specimens were entirely flat during the measurements, therefore, a 2D image analysis was sufficient for revealing the SR of the specimens.

2.8 Thermogravimetric Analysis

A thermogravimetric analyzer (TGA Q5000, TA Instruments, USA) was employed to explore the thermal stability of PET melt-spun fibres. The measurement was conducted in a nitrogen atmosphere, where the scan heating rate was 10 °C/min with an investigated range from 25 to 800 °C.

2.9 Size Exclusion Chromatography

In order to reveal the weight-average molar mass M_w, number-average molar mass M_n, and polydispersity M_w/M_n of all the investigated materials, size exclusion measurements (GPCmax, Malvern) with triple detection were conducted. The PET was dissolved in HFIP as solvent at 50 °C for 2 h at a concentration of 2 g/L, in order to provide a complete dissolution of PET before conducting the measurements. With a flow rate of 0.8 mL/min, the GPC measurements were performed at a temperature of 35 °C.

Melt-spun fibres subjected to a higher processing temperature of 280 °C were analyzed as well to evaluate the thermal degradation behaviour.

2.10 Rheological Characterization

The rheological properties of all the PET materials in oscillatory shear were investigated using an ARES rheometer (TA Instruments, USA) in the plate-plate geometry. Through this type of rheometer may only be available at a low angular frequency (<100 rad/s), it still can provide the critical rheological information corresponding to the polymer molecular structure [41]. The dried PET materials were previously hot compression at 270 °C under vacuum to produce cylindrical specimens with 2 mm in thickness and 25 mm in diameter. For all the prepared specimens, dynamic mechanical measurements in the air were performed at 270 and 280 °C, respectively, at the frequency ranging from 100 to 0.1 rad/s. A deformation amplitude of 10% in the linear viscoelastic zone proved previously by a strain amplitude sweep measurement, was used for all the specimens in this study.

3 Characterizations of Raw Materials

3.1 Thermal Behaviour

The DSC thermograms of all the used PET materials were plotted in Fig. 2, where the different coloured flakes from ***rPET-A*** and ***rPET-B*** were chosen to analyze the information comprehensively. As shown, obvious double melting peaks (marked as T_{m1} and T_{m2}) were observed for the coloured flakes, which suggested that the

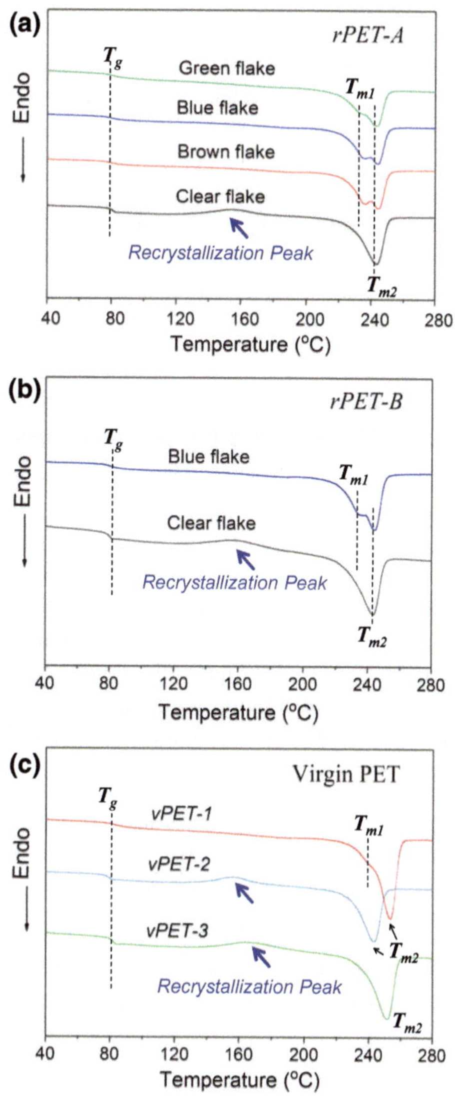

Fig. 2 DSC traces of **a** ***rPET-A***, **b** ***rPET-B***, **c** virgin PET (***vPET-1***, ***vPET-2*** and ***vPET-3***) obtained from the DSC second heating scan. Redrawn by authors from Ref. [12]

addition of pigment could influence the crystallization behaviour of PET to some degree. The data derived from the second heating scan of all the investigated PET materials are summarized in Table 4. As reported that the glass transition temperature T_g, especially the melting temperature T_{m2} of ***vPET-1*** and ***vPET-3*** were higher than those of other PET polymers, which was expected due to the difference between homo-polymers (***vPET-1*** and ***vPET-3***) and co-polymers. This is because the random sequence of the co-monomeric units in co-polymers would destroy the crystalline lattice compared with the homo-polymers, and thus less ordered, unstable structures are formed [42]. Consequently, the disordered and unstable structures lead to the lower melting temperature of co-polymers.

It can also be found that the first melting peak ($\boldsymbol{T_{m1}}$) is about 10 °C lower than the second melting peak ($\boldsymbol{T_{m2}}$), where the appearance of $\boldsymbol{T_{m1}}$ is documented to be more likely concerning to the lower lamellae thickness and more imperfection of crystals [43]. Instead, the appearance of a single sharp melting peak in the thermograms, for instance, ***vPET-3*** in Fig. 2c, indicates comparatively perfect crystalline structures and a higher lamellae thickness. In addition, with respect to some of the specimens, like the clear flake in ***rPET-A***, a recrystallization peak can be observed at 153–165 °C. The crystallinity X_c was calculated by Eq. (7) and listed in Table 5 as well. Overall, the crystallinity X_c of all the materials is lower than 25%. There was a slight increase of crystallinity X_c in coloured flakes compared with the clear ones in ***rPET-A*** and ***rPET-B***.

3.2 *Molar Mass Characterization*

The corresponding weight-average molar mass M_w, number-average molar mass M_n, and polydispersity M_w/M_n of the as-received PET raw materials are given in Table 6. As shown, ***vPET-3*** has the highest number- and weight-average molar mass, while ***vPET-1*** has the lowest molar mass. For the recycled PET, ***rPET-B*** (the purer one) shows a slightly higher molar mass than that of ***rPET-A***. Furthermore, ***rPET-B*** has the narrowest molar mass distribution and ***vPET-2*** has the broadest. The molecular fingerprint information acting as the most crucial factors, would remarkably influence the spinnability of each material in the following fibre-spinning process.

Table 6 Weight-average molar mass M_w, number-average molar mass M_n, and polydispersity M_w/M_n of PET raw materials

	Raw materials				
	rPET-A	*rPET-B*	*vPET-1*	*vPET-2*	*vPET-3*
$\boldsymbol{M_n}$ (kg/mol)	29.3 ± 4.2	32.3 ± 0.8	23.3 ± 1.4	25.5 ± 2.4	34.4 ± 0.8
$\boldsymbol{M_w}$ (kg/mol)	53.6 ± 0.6	56.9 ± 0.3	42.9 ± 0.9	50.4 ± 4.1	62.5 ± 5.9
$\boldsymbol{M_w/M_n}$	1.8 ± 0.2	1.7 ± 0.1	1.8 ± 0.1	2.0 ± 0.3	1.8 ± 0.2

3.3 Rheological Characterization

Figure 3 shows the complex viscosity as a function of the frequency for all analyzed specimens at two temperatures. It clearly indicates that at 270 and 280 °C, the viscosities of ***rPET-A***, ***rPET-B***, ***vPET-1***, ***vPET-2*** are almost independent of frequency, showing the Newtonian plateau within the whole frequency range examined (0.1–100 rad/s). Only ***vPET-3*** suggests a weak shear thinning behaviour at high frequency, which should be attributed to its higher molar mass considering that the higher molar mass shows more sensitive to the shear field [44]. The Cox-Merz rule, i.e., the complex viscosity $|\eta^*|(\omega) = \eta(\dot{\gamma})$ for $\omega \equiv \dot{\gamma}$ [45], is employed to convert complex viscosity $|\eta^*|$ as a function of the angular frequency ω, into the shear viscosity η as a function of the shear rate $\dot{\gamma}$. Then the Carreau-type function [46], Eq. (9), was introduced to obtain the zero shear viscosity η_0 by extrapolation for all the materials. It should be noted that due to the limitation of the investigated frequency, only η_0 is accessible. The extrapolated values of η_0 are provided in Table 7.

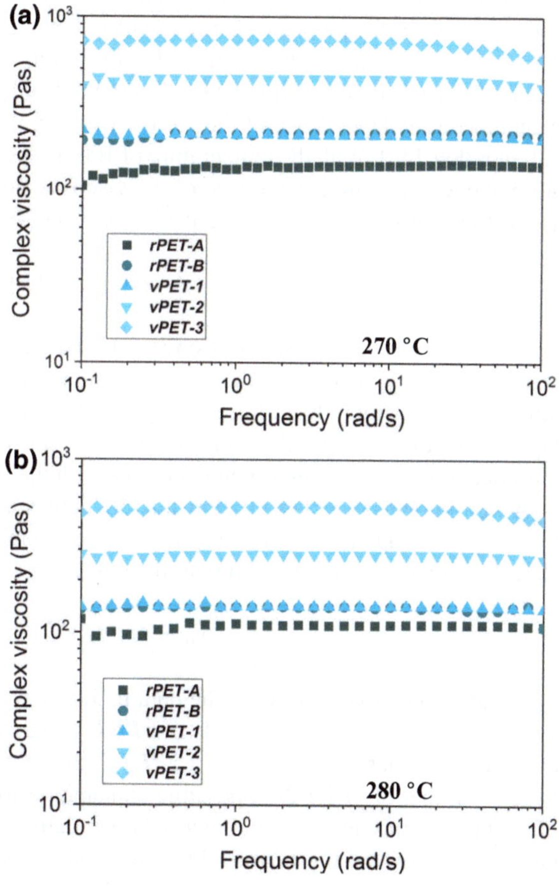

Fig. 3 Complex viscosity versus frequency for all the PET specimens at **a** 270 °C and **b** 280 °C

Table 7 Zero shear viscosity η_0 of each material at two different temperatures

	Zero shear viscosity η_0 (Pa s)				
	rPET-A	***rPET-B***	***vPET-1***	***vPET-2***	***vPET-3***
270 °C	148.3 ± 3.1	195.4 ± 1.3	195.2 ± 1.5	401.3 ± 4.1	744.5 ± 7.5
280 °C	115.3 ± 1.7	146.7 ± 5.5	130.3 ± 1.5	301.8 ± 3.4	535.3 ± 4.2

$$\left|\eta^*\right| = \eta_0/(1 + (\omega/\omega_c)^2)^{m/2} \tag{9}$$

where η_0 refers to the zero shear viscosity, ω_c is the critical angular frequency. Namely, the complex viscosity function begins to decline once the angular frequency exceeds the ω_c. m is a power index.

From the data in this chart, a strong decrease of η_0 can be observed in all the specimens with the increasing temperature from 270 to 280 °C. The reasons can be revealed in two aspects: (i) the increasing mobility of the chains at a higher temperature, and (ii) the potential thermal oxidative degradation of the molecular chain. As known, chain entanglement, interchain friction, and free volume are important factors determining the rheological properties of polymers [47]. In polymer melts, the temperature dependence of the viscosity is one of the most significant factors in polymer flow [48]. With the temperature of 100 °C above the glass transition temperature, the temperature dependence of the viscosity polymer melts can be presented as the Arrhenius-type Equation derived by Eyring [49],

$$\eta_0 = A \cdot \exp\left(E^*/RT\right) \tag{10}$$

where η_0 denotes the zero shear viscosity, A refers to a constant, E^* denotes the flow activation energy, R refers to the gas constant, and T denotes the absolute temperature. E^* varies widely for different polymer, depending on chain composition. The higher activation energy, the more temperature sensitive is the melt. According to this equation, η_0 of polymer melt decreases with the increasing temperature.

A well-known power law relationship [50] between the zero shear viscosity η_0 and the weight-average molar mass M_w was applied in this study, as shown in Eq. (11). Here U is a parameter depending on the temperature.

$$\eta_0 = U \cdot {M_w}^{3.4} \tag{11}$$

Make a comparison between the data in Tables 6 and 7, it can be seen that even though the recycled PET materials of ***rPET-A*** and ***rPET-B*** have a higher weight-average molar mass M_w than that of ***vPET-2***, the zero shear viscosity η_0 of them are lower than that of ***vPET-2***. A possible explanation for this should be the existence of contaminations and potential thermal degradation of the recycled PET materials. Given this, only the three virgin PET materials at two different temperature were evaluated, and the double-log plot of the zero shear viscosity η_0 as a function of the weight-average molar mass M_w is shown in Fig. 4. The fitted lines are the best fit using Eq. (11), which present good agreements with the measured data points.

Figures 5 and 6 show the storage modulus G' and loss modulus G'' as a function of frequency ω for five materials at two experimental temperatures. As expected, all the materials indicate the linear molecular chain, considering that the molecular chains are completely relaxed and present characteristic behaviour in the terminal zone, where G' is proportional to ω^2 and G'' proportional to ω [48]. Additionally, the relationship between the microstructure of polymers or blends and the slope in the plot of storage modulus G' and loss modulus G'' has been investigated by Han et al. [51–53]. As acceptable for low frequency even if applying a simply Maxwell-model for first approximation, generally the majority of isotropic homogeneous polymer melts and solutions show a slope of 2 in the logarithmic plot of G' against G'', which corresponds well to this work. It should be pointed out that due to the resolution of the rheometer used, data points of G' were dominated by the noise at a lower frequency.

It can also be observed that the storage modulus G' and loss modulus G'' at a lower temperature of 270 °C in Fig. 5 presents the same trend with the zero shear viscosity η_0 as provided in Table 7. For ***rPET-B*** and ***vPET-1*** with the similar zero shear viscosity η_0, they also have almost exactly the same value of storage modulus G' and loss modulus G'' at the same frequency. With respect to the storage modulus G' and loss modulus G'' at the higher temperature of 280 °C, the curves of ***vPET-2*** and ***vPET-3*** are coinciding; simultaneously, the curves of ***rPET-B***, ***vPET-1*** and ***rPET-A*** are overlapping considerably (see Fig. 6). That is to say, at higher temperatures, the difference of the storage modulus G' and loss modulus G'' between different materials becomes insignificant. Furthermore, with the increase of temperature, a slight drop of the storage modulus G', as well as the loss modulus G'' of all the materials is also observed as shown in Fig. 7.

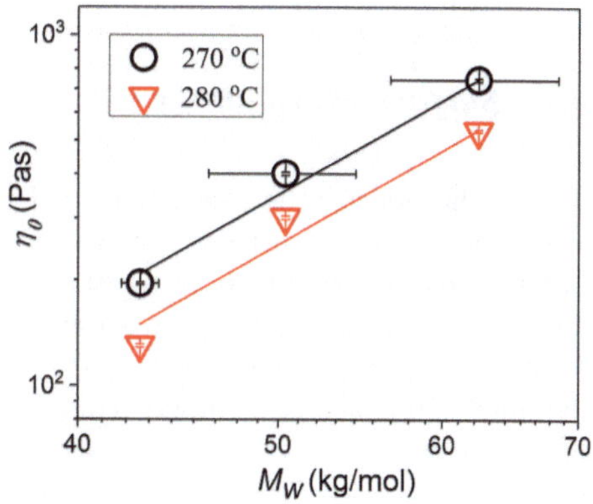

Fig. 4 Zero shear viscosity η_0 versus weight-average molar mass M_w for three virgin PET materials at two temperatures. The lines are the best fit using Eq. (11)

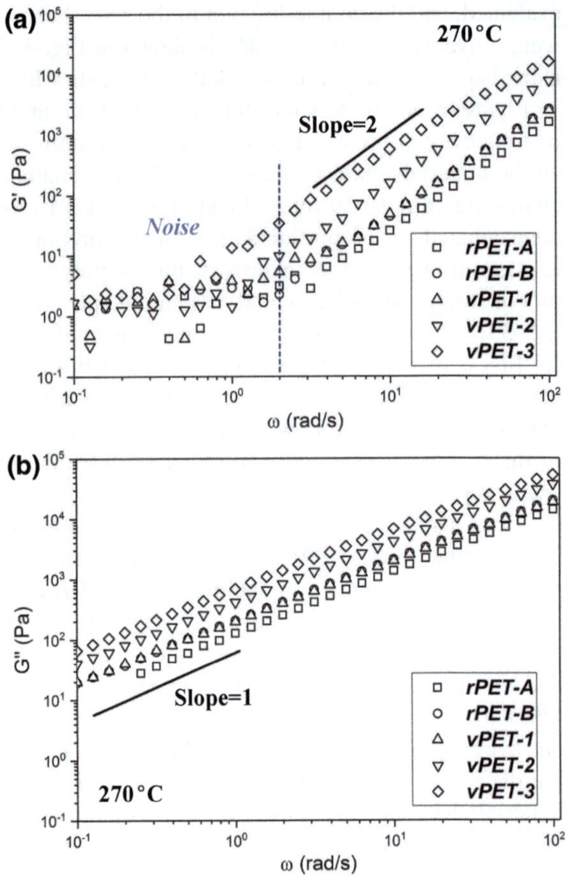

Fig. 5 **a** Storage modulus G' and **b** loss modulus G'' versus frequency ω for all the PET specimens at 270 °C

4 Characterizations of Melt-Spun Fibres

4.1 Surface Morphology

To meet the industrial demand of the textile applications, the polymer fibres produced shall have a smooth surface. The surface smoothness of the PET melt-spun fibres in this study can be influenced by (i) the instabilities of PET melt leaving from the die (negative factor) and (ii) the aerodynamic stretching via the aspirator (positive factor). Therefore, to better focus on the effect of inherent instabilities of PET melts, the fibres fabricated at the lower temperature of 270 °C combined with the lowest take-up pressure of 0.5 bar were chosen for SEM observation. To clearly observe the surface topography, SEM images are shown in Fig. 8 with standard SEM imaging (left) and contrast improved imaging (right). Though the diameter of fibres from different materials varies significantly, all the fibres apart from the one produced

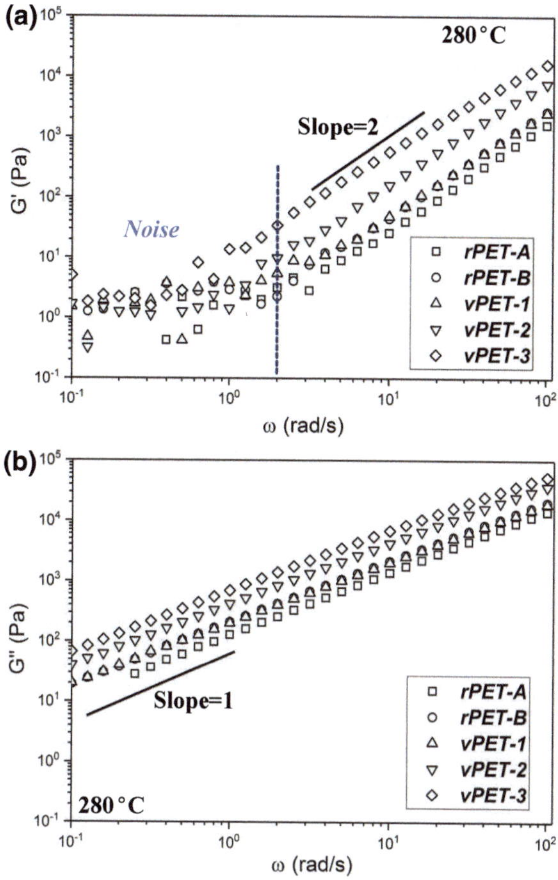

Fig. 6 **a** Storage modulus G' and **b** loss modulus G'' versus frequency ω for all the PET specimens at 280 °C

by ***vPET-3*** have a smooth surface. The rough surface in ***vPET-3*** fibre is marked by yellow arrows. Generally, a rough surface indicates the an instability of postulated melt during the melt spinning process, which may due to the higher melt viscosity, i.e., poor flowability of ***vPET-3*** melt (as shown in Table 7, its zero shear viscosity η_0 is strikingly higher than others). The melt instability of ***vPET-3*** suggests the limitation of its spinnability, which will be revealed in the following section.

4.2 Diameter Distribution of Fibres

In this work, six take-up pressures from 0.5 to 3.0 bar with 0.5 bar increments were applied to stretch the PET filaments leaving from the die. However, it was found that not all the investigated take-up pressures were suited for each PET material. Figure 9 presents the selected pictures of the PET fibres drawn at the lowest (0.5 bar)

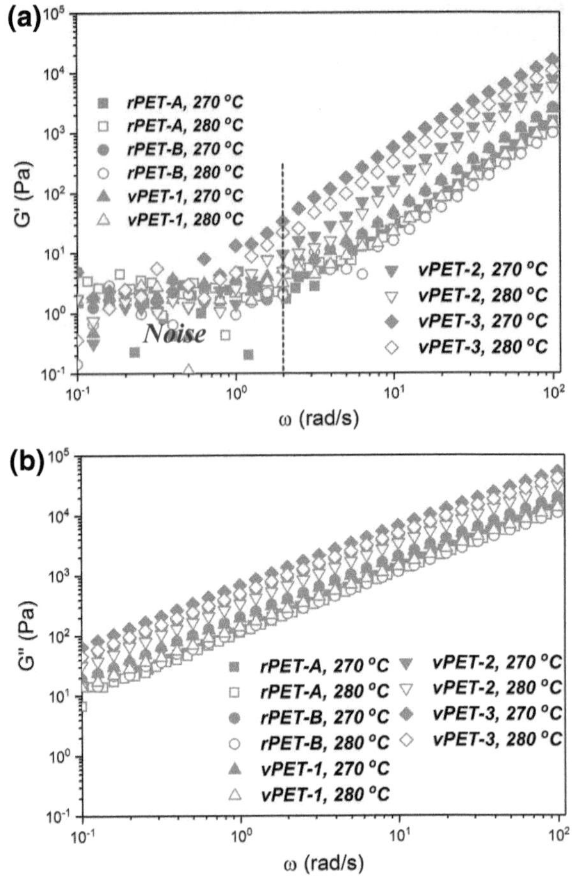

Fig. 7 **a** Storage modulus G' and **b** loss modulus G'' versus frequency ω for all the PET specimens at different temperatures

and highest take-up pressure (3.0 bar) at two processing temperatures. In terms of ***rPET-A***, ***rPET-B*** and ***vPET-1***, a uniform fibre diameter distribution can be observed under each experimental condition, which should result from their optimal molar mass, molar mass distribution (see Table 6, $M_w/M_n \approx 1.8$) and melt viscosity for fibre spinning (Table 7).

One can observe that for the ***vPET-2*** extruded at 280 °C and ***vPET-3*** extruded at 270 °C, the maximal available take-up pressure was limited at 2.0 bar. The fibre diameter distribution becomes broader with the increasing take-up pressure for ***vPET-2*** and ***vPET-3***, until fibres eventually break. This broader diameter distribution is exemplified in Fig. 10 using one microscopic picture from ***vPET-3*** at 270 °C and 2.5 bar. As mentioned above, the spinnability of ***vPET-3*** at 270 °C was restricted by its melt instability on account of the highest melt viscosity. However, the high melt viscosity could not be used to explain the deterioration of spinnability for ***vPET-2*** at 280 °C, owing to that ***vPET-2*** at 270 °C presents a desirable spinnability in spite of

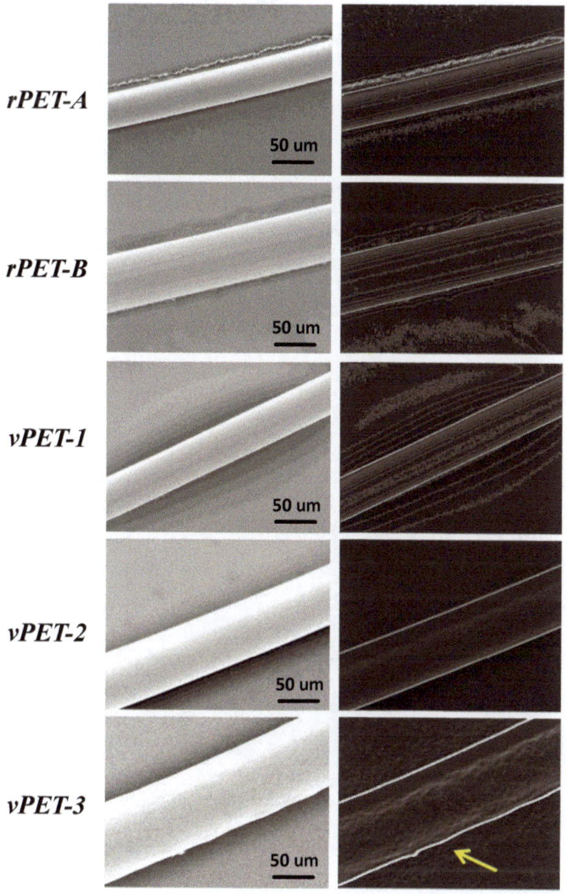

Fig. 8 SEM images of PET fibres produced at 270 °C, 0.5 bar. The contrast and brightness on the edges of fibres are increased, before (left) and after (right)

a higher melt viscosity than that at 280 °C. A potential explanation for this might be the broader molar mass distribution of ***vPET-2***, with $M_w/M_n \approx 2.0$ in Table 6.

To quantitatively study the fibre diameter at two processing temperatures, an average of fibre diameter (at least 30 fibres were randomly chosen and measured), as well as a standard deviation for each material versus the applied take-up pressure, is presented in Fig. 11. The relative length of the error bar in this plot (i.e., the deviation of the presented fibre diameter) indicates the inhomogeneity of the PET fibres diameter. Thus, the variations of error bar provide convincing evidence that the melt-spun fibres from ***rPET-A***, ***rPET-B*** and ***vPET-1*** have more uniform diameter than those from ***vPET-2*** and ***vPET-3***, which is consistent with the fibre morphology in Figs. 9 and 10. For melt-spun fibres from ***vPET-2*** at 280 °C (green spheres) and melt-spun fibres from ***vPET-3*** at 270 °C (blue cubes), it must be pointed out that the larger error bars refer to a poor spinnability of both materials at such processing condition, even though the melt-spun fibres can be produced continuously at below 2.0 bar.

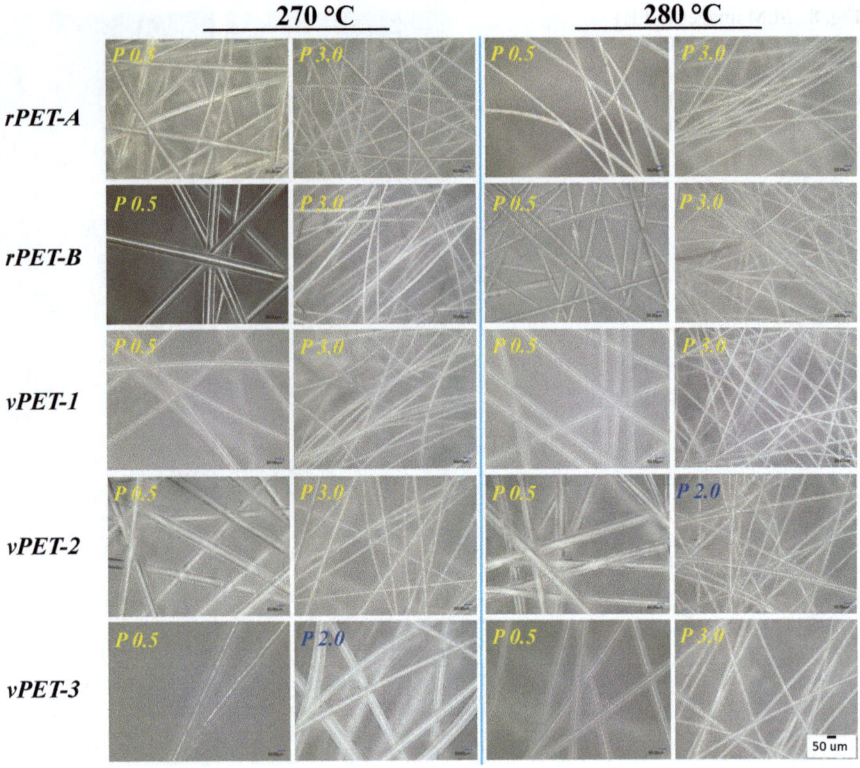

Fig. 9 Melt-spun fibres produced under the lowest the highest viable take-up pressures at two processing temperatures

Furthermore, the following findings can also be summarized. (i) With respect to all the investigated PET materials, the diameter of the PET fibres decreases as the applied take-up pressure increases, which is valid for both processing temperatures. This result can be explained by Eq. (4): as the throughput Q keeps constant, the diameter of the fibres D_F decreases as the v_F increases. (ii) When the applied take-up pressures are below 2.0 bar, the diameter of the PET fibres extruded at 270 °C is generally larger than those extruded at 280 °C. Such a phenomenon can be thus explained by considering the concept of rheology: a higher processing temperature leads to a lower melt viscosity, which means, a better flowability. A better flowability of the PET melt ensures the extruded filaments are easily drawn and yield finer fibres. (iii) As the applied take-up pressure in the aspirator increases, the difference in the diameter of the fibres produced at two processing temperature reduces, which is valid for all the materials. An increased take-up pressure leads to faster air flow in the aspirator, therefore, the melt-spun fibres would be cooled down much faster. The increased cooling rate leads to a rapid solidification of the PET fibres, and as a consequence, the draw ratio will be restricted to a great extent.

Fig. 10 Overview of ***vPET-3*** fibres generated at 270 °C and 2.5 bar, indicating an inhomogeneity of the fibres diameter

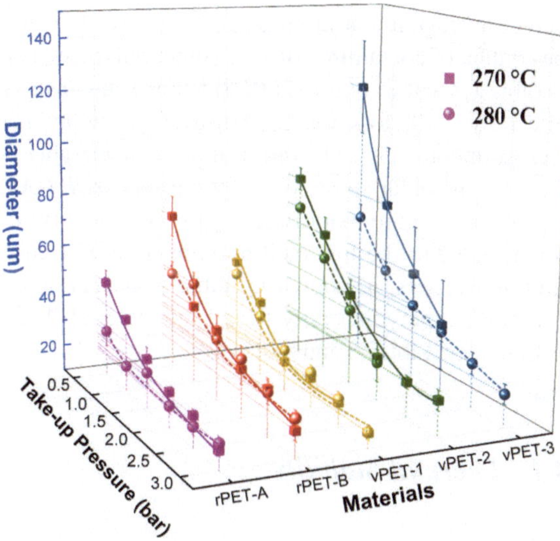

Fig. 11 Average diameter of the PET fibres with standard deviation versus the take-up pressure at two processing temperatures of 270 and 280 °C. The lines are a guide to the eyes. Redrawn by authors from Ref. [12]

Table 8 Weight-average molar mass, number-average molar mass, polydispersity of the melt-spun fibres at 280 °C

	Melt-spun fibres				
	rPET-A	***rPET-B***	***vPET-1***	***vPET-2***	***vPET-3***
M_n	21.5 ± 1.5	26.6 ± 2.6	22.5 ± 0.9	21.4 ± 0.4	29.9 ± 2.9
M_w	37.5 ± 0.2	45.4 ± 1.5	40.2 ± 1.1	40.8 ± 1.1	51.3 ± 2.1
M_w/M_n	1.8 ± 0.1	1.7 ± 0.2	1.8 ± 0.1	1.9 ± 0.1	1.7 ± 0.2

4.3 Molar Mass Degradation

Thermal degradation, as an important issue in polymer processing, should not be ignored. As shown in Table 8, the weight-average molar mass M_w, number-average molar mass M_n, and polydispersity M_w/M_n of all the PET melt-spun fibres fabricated at the higher temperature of 280 °C, at 0.5 bar were measured to compare with the data of raw materials in Table 6. To make a clear comparison of the molar mass variation before and after the fibre spinning process at 280 °C, the histograms combined with the calculated corresponding molar mass loss values are plotted in Fig. 12.

The molar mass loss data apparently reveal that ***rPET-A*** experienced a large degree of thermal degradation during melt spinning process. Its loss of number and weight-average molar mass are respectively 26.6 and 30.0%, which originates from the content of contamination (e.g., other polymers, paper label and so forth) as shown in Table 1. On the contrary, ***vPET-1*** shows the lowest degree of thermal degradation, with only 3.4% loss for number-average molar mass and 6.3% loss for weight-average molar mass. Overall, a more considerable degree of thermal degradation, i.e., loss of molar mass, of co-polyesters (***rPET-A***, ***rPET-B*** and ***vPET-2***) can be found compared to that of the homo-polyesters (***vPET-1*** and ***vPET-3***). This should be attributed to the low melting temperature of co-polyesters compared to that of homo-polyesters, as shown in Table 5. Remarkably, no obvious differences were found in the polydispersity for all the materials before and after the melt spinning process (see Tables 6 and 8).

4.4 Thermal Stability

In the previous study, a significant difference (>10 °C) in decomposition temperature between recycled PET fibres and virgin PET fibres has been reported [54]. Similarly, in this work, the difference in decomposition temperature of all the PET fibres also lies in the range of 10 °C. As shown in Fig. 13. ***rPET-A*** has the lowest decomposition temperature, at about 365 °C, and ***vPET-2*** presents the highest one, at about 375 °C. This should be easily understood considering that other impurities in ***rPET-A*** with lower thermal stability start decomposing at a lower temperature. In addition, the residual weight fraction of all the fibres are also in a small range of 10–13%, that

Fig. 12 Comparison of the **a** number- and **b** weight-average molar mass between raw materials and fibres produced at 280 °C. The molar mass loss of the fibres subjected to the melt spinning process is shown as well

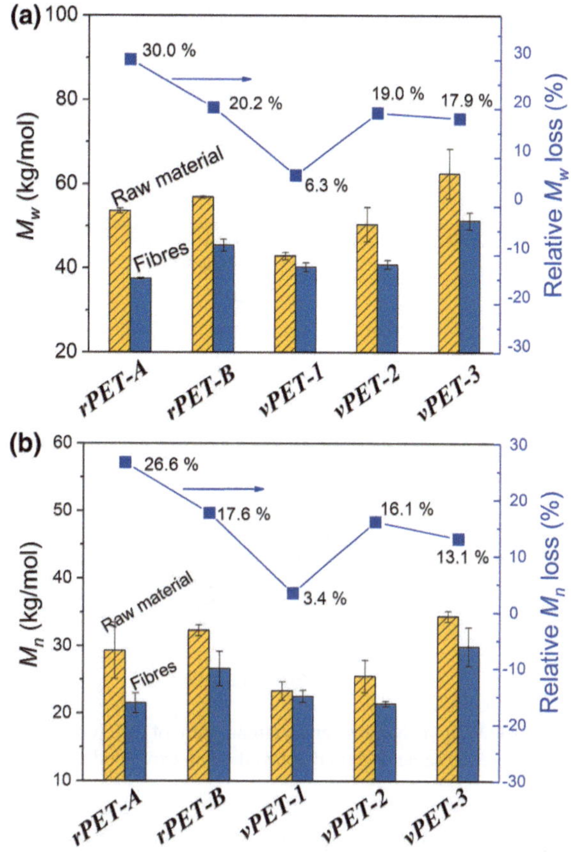

is, the fibres produced from virgin PET show no obvious inferior thermal stability in comparison with the ones from recycled PET. Interestingly, both the highest and lowest residual weight of fibres come from the recycled PET materials, i.e., ***rPET-B*** has the highest residual weight and ***rPET-A*** has the lowest one. A potential explanation for the result may be also the existence of uncertain contaminants in the recycled PET: some decompose at low temperatures, while some of them are stable even at 800 °C.

4.5 Evaluation of Mechanical Properties

4.5.1 Investigation of Gauge Length and Test Speed

As known, the measured mechanical properties of products would be influenced by the environmental condition, such as temperature and humidity, and the experimental

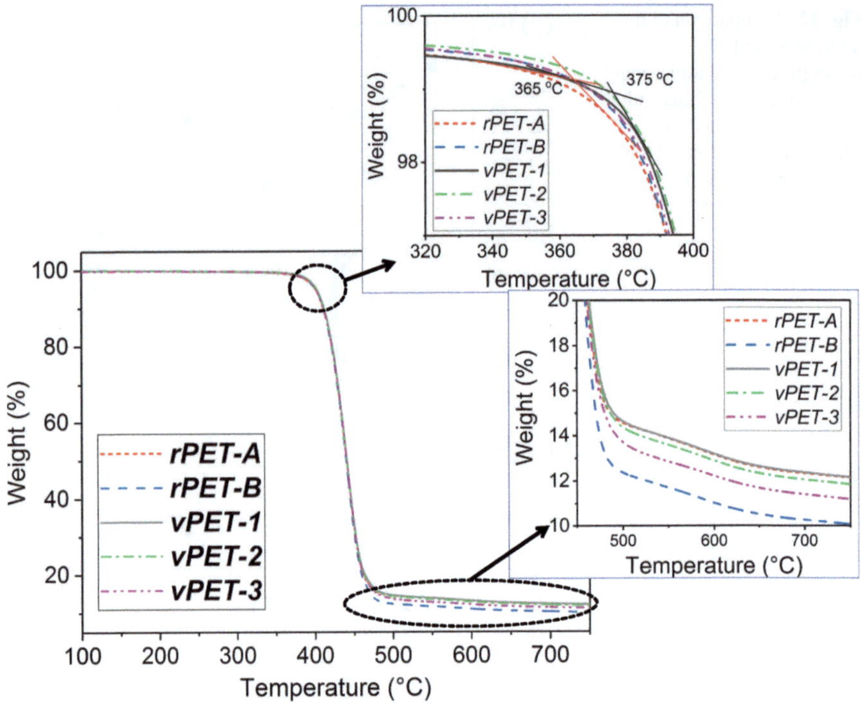

Fig. 13 Weight changes versus temperature of different fibres produced at 270 °C and 0.5 bar. The insets are the same spectra displayed on an expanded axis scale

parameters, mainly referring to the gauge length and test speed used. To clearly define the possible effect of different experimental parameters on the measured force at break and elongation at break in this study, one kind of fibres from ***vPET-1***, the fibre-grade virgin PET without other contaminations, generated at 270 °C, 1.0 bar was selected because of the stable structure and uniform diameter. Considering the displacement range of the clamps in the used equipment, three effective gauge length of 5, 10, 15 mm combined with different test speeds of 50, 100, 150, 200 and 300 mm/min were employed in measurements at room temperature. For each combination, at least 20 fibres were measured to stretch until failure.

With the gauge length of 5 mm, Fig. 14 illustrates that the average values of force at break and elongation at break increase slowly with the increase of test speed from 50 to 300 mm/min. For elongation at break, the average values increased slowly and remained almost constant at high test speeds (i.e., 200 and 300 mm/min). When the gauge length was adjusted to be 10 mm, in Fig. 15, no significant differences in the average values of force at break, and the elongation at break with the increasing test speed are found. All the data points obtained at different test speeds with the gauge length of 10 mm are provided as well in Fig. 16. However, based on the equivalent Y-scale in Figs. 14 and 15, a reduction in elongation at break values can be observed

Fig. 14 **a** Force at break and **b** elongation at break versus test speed of ***vPET-1*** fibres (270 °C, 1.0 bar) with the gauge length of 5 mm

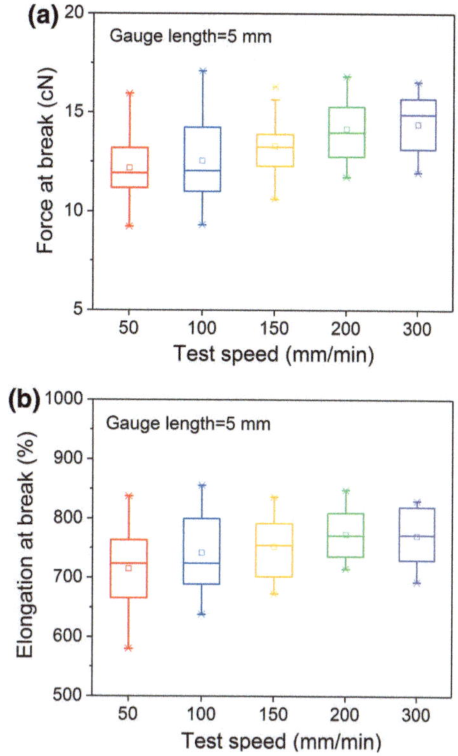

when the gauge length increased from 5 to 10 mm. Therefore, as discussed above, the gauge length plays a more important role than the test speed in the measured results of mechanical properties.

In Fig. 17, all the data points obtained from different gauge length (5, 10 and 15 mm) at the constant test speed of 100 mm/min are plotted. From the data cloud area, it is clearly found that the gauge length of the specimen can affect the measured elongation at break. That is to say, generally the larger gauge length induces the lower elongation at break values, which is in agreement with the 'weakest link' theory [55]. Nevertheless, the plot statistically proves that the gauge length has a slight influence on the measured force at break values.

4.5.2 Mechanical Properties of the Melt-Spun PET Fibres

To measure the tenacity and elongation at break of all the PET fibres, the gauge length of 10 mm and the constant test speed of 100 mm/min were selected. At least 30 fibre specimens were stretched to failure for each group. Figure 18 presents the tenacity of the melt-spun PET fibres at two processing temperatures versus the applied take-up

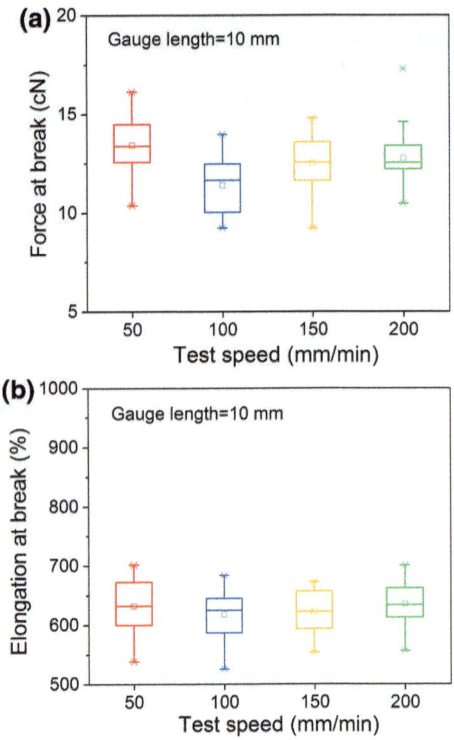

Fig. 15 **a** Force at break and **b** elongation at break versus test speed of ***vPET-1*** fibres (270 °C, 1.0 bar) with the gauge length of 10 mm

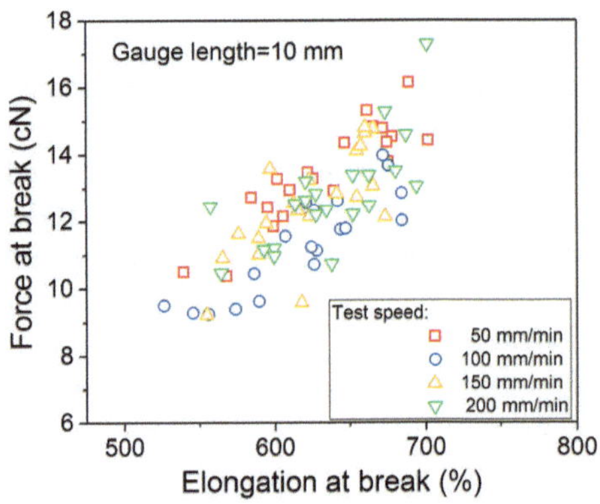

Fig. 16 Data points of force at break versus elongation at break of ***vPET-1*** fibres (270 °C, 1.0 bar) at different test speeds with the gauge length of 10 mm

pressure in the aspirator. It can be seen that the tenacity of all the investigated PET fibres increases as the applied take-up pressure increases, which is valid for both processing temperatures.

For the PET fibres produced at a lower take-up pressure (below 2.0 bar), the tenacity presents a gradual increase as the applied take-up pressure increases, and the fibres produced from virgin PET grades show no significant difference comparing with that from the recycled PET. For the PET fibres produced at a higher take-up

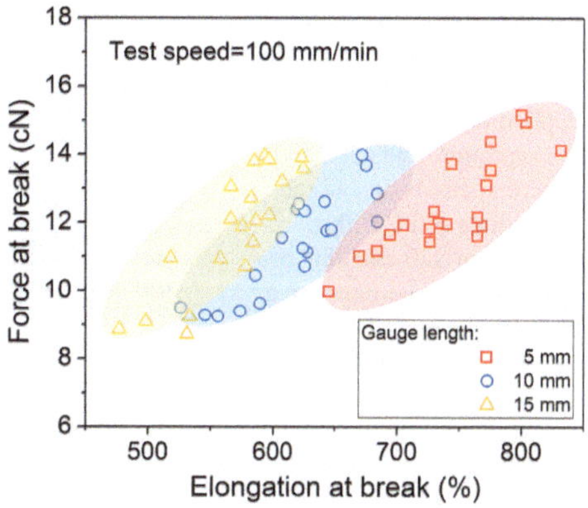

Fig. 17 Force at break versus elongation at break of ***vPET-1*** fibres (270 °C, 1.0 bar) at a test speed of 100 mm/min with different gauge length

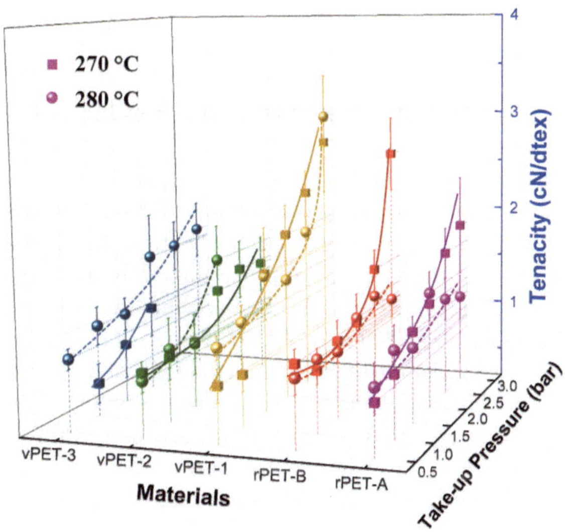

Fig. 18 Tenacity of all the investigated PET fibres versus the applied take-up pressure in the aspirator. The cube and sphere symbols denote the data points of fibre extruded at 270 and 280 °C, respectively. The lines are a guide to the eyes. Redrawn by authors from Ref. [12]

pressure, the ***vPET-1*** melt-spun fibres display the optimal tenacity comparing with other virgin PET grades. Concerning the recycled PET, ***rPET-B*** melt-spun fibres achieve its maximal tenacity at 270 °C, 3.0 bar, which indicates its obvious superiority compared with ***rPET-A*** fibres. Additionally, this maximal tenacity is even comparable to that of ***vPET-1***.

In terms of the elongation at break, as shown in Fig. 19, the opposite trend compared to the tenacity can be clearly observed. Namely, as the take-up pressure increases, the elongation at break of all the measured PET fibres quickly drops, which is valid for both processing temperatures. Furthermore, it should also be noted that ***rPET-A*** melt-spun fibres have the lowest elongation at break compared with all the other materials.

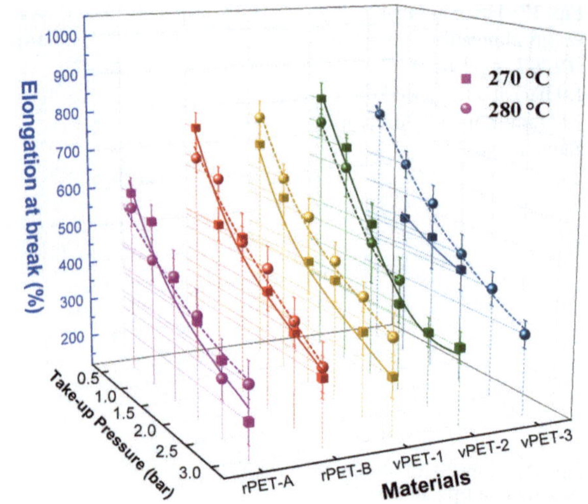

Fig. 19 Elongation at break of all the investigated melt-spun fibres versus the applied take-up pressure in the aspirator. The cube and sphere symbols denote the data points of PET fibre extruded at 270 and 280 °C, respectively. The lines are a guide to the eyes. Redrawn by authors from Ref. [12]

4.5.3 Crystallinity and Orientation of Fibres from *rPET-B* and *vPET-1*

Based on the desirable properties, e.g., spinnability, tenacity and elongation at break of the melt-spun fibres, ***rPET-B*** and ***vPET-1*** should be defined as the optimal materials for melt spinning respectively from the recycled and virgin PET. It must be noted that the mechanical properties of the melt-spun fibres highly depend on the crystallinity and molecular chain orientation [16, 59–62]. Therefore, the fibres fabricated from ***rPET-B*** and ***vPET-1*** were selected to clarify the impacts of crystallinity and molecular chain orientation via DSC and hot shrinkage measurements.

For simplicity, the melt-spun fibres with the highest tenacity, i.e., ***rPET-B*** fibres extruded at 270 °C, drawn at 3.0 bar, and ***vPET-1*** extruded at 280 °C, drawn at 3.0 bar, were investigated. The fibres collected at the same processing temperature, but drawn at a lower take-up pressure of 1.0 bar were applied to perform a corresponding comparison with each of them. Figure 20 presents the first heating scan

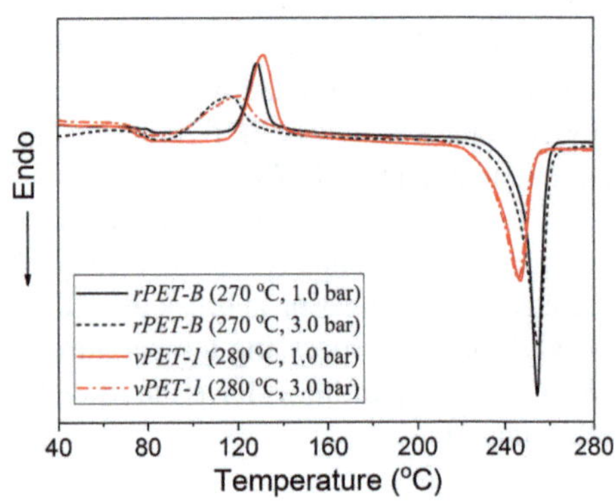

Fig. 20 Thermograms via DSC first heating scan of the selected ***rPET-B*** and ***vPET-1*** fibres

Table 9 DSC data derived from Fig. 20 and shrinkage ratio (SR) of the selected ***rPET-B*** and ***vPET-1*** fibres

Fibres	T_g (°C)	$T_{c,onset}$ (°C)	T_c (°C)	ΔH_c (J/g)	T_m (°C)	ΔH_m (J/g)	X_c (%)	SR (%)
rPET-B (270 °C, 1.0 bar)	74.6	120.1	131.6	23.3	247.0	35.5	9.0	10.0 ± 4.3
rPET-B (270 °C, 3.0 bar)	74.2	95.1	120.8	21.2	246.4	36.7	11.4	51.6 ± 9.3
vPET-1 (280 °C, 1.0 bar)	81.1	120.7	128.7	15.4	254.4	41.4	19.1	11.5 ± 2.3
vPET-1 (280 °C, 3.0 bar)	80.1	94.4	116.9	20.2	254.9	43.6	17.2	39.3 ± 5.5

Rearranged by authors from Ref. [12]

of the measured PET fibres from the DSC thermograms. Table 9 summarizes the derived data from the thermograms, and hot shrinkage ratio SR was calculated via Eq. (8). Comparing with the data in Table 5 of the PET raw materials, it can be revealed that the melt-spun fibres always have low crystallinity X_c, which is a result of the rapid cooling speed during the fibre stretching procedure in the aspirator.

Interestingly, the crystallinity X_c of the melt-spun fibres produced from the same PET material presents similar results independent of the applied take-up pressure. In addition, SR of the ***rPET-B*** fibres generated at 270 °C, 1.0 bar and ***vPET-1*** fibres generated at 280 °C, 1.0 bar is almost equivalent. However, the crystallinity X_c of ***vPET-1*** (homo-polyester) is approximately double than that of ***rPET-B*** (co-polyester). It should be the presumable reason for a higher tenacity of ***vPET-1*** fibres produced at 280 °C, 1.0 bar than that of ***rPET-B*** fibres produced at 270 °C, 1.0 bar, in Fig. 18.

As shown in Fig. 11, it can be seen that as the take-up pressure normally increases, the fibre diameter decreases, and thus an increasing draw ratio DR can be induced (see Eq. (6)). Meanwhile, the draw ratio DR has been proved to have a positive influence on the inner molecular chain orientation along the fiber axis of the melt-spun fibres [56, 57]. Furthermore, a higher shrinkage ratio SR of fibres exactly reflects a higher inner molecular chain orientation, which consequently strengthens the fibres [16, 31, 56–58]. This is consistent with the finding that tenacity grows with the increasing take-up pressure in Fig. 18. On the contrary, the increased molecular chain orientation would result in a decline of the elongation at break [10], in agreement with the results in Fig. 19.

More importantly, it can be concluded that the similar mechanical properties of ***rPET-B*** fibres extruded at 270 °C, drawn at 3.0 bar, and ***vPET-1*** fibres extruded at 280 °C, drawn at 3.0 bar, originate from different factors. With respect to ***rPET-B*** fibres, the molecular chain orientation rather than the crystallinity plays a leading role, because of a higher SR. This conclusion can also be supported by the lower tenacity of ***rPET-B*** fibres produced at a higher processing temperature of 280 °C, drawn at 3.0 bar, in Fig. 18, where more oriented molecular chains are relaxed during the longer cooling time. While regarding ***vPET-1*** fibres, its mechanical properties mainly depend on its higher crystallinity X_c, together with the secondary influence of increased inner molecular chain orientation.

5 Performance-Reliability Plot

Weibull distribution, normally applied to analyze the durability, can provide a statistical method to present the performance and reliability of specimens. This has been successfully applied to mechanical performance in the field of polymeric materials [59, 60]. With regard to the stress at break, the Weibull distribution can be written as

$$P = 1 - \exp[-(\sigma_b/\sigma_{b,0})^m] \tag{12}$$

where P means the cumulative probability of failure, σ_b refers to the stress at break, $\sigma_{b,0}$ refers to the characteristic stress at break where 63.2% of the specimens break, and the Weibull modulus m can be obtained to evaluate the width of the distribution. When the $\sigma_{b,0}$ keeps constant, an increased Weibull modulus m corresponds to an increased slope of the cumulative probability curve, mirroring a remarkable reproducibility, i.e., better reliability.

In order to assess the test group by the Weibull distribution, all the performance values of specimens need to be ordered, such as $\sigma_1 \leq \sigma_2 \leq \cdots \sigma_i \cdots \leq \sigma_n$. Afterwards, a probability P_i is successively distributed to each value of stress at break, where P_i is obtained using Eq. (13), based on the Bergman method [61]. This would offer the lowest coefficient of variation of m and any Weibull parameter x_0 for a total number of n specimens. The Weibull modulus m and the characteristic Weibull parameter x_0 can be revealed by Eq. (12).

$$P_i = (i - 0.5)/n \tag{13}$$

As an important industrial parameter showing the distribution of the material characteristics, the Weibull moduli m of different measured properties from ***rPET-B*** and ***vPET-1*** fibres were exemplified to explore their performance and reliability. The highest values of Weibull moduli m represent the narrowest distribution of all the measured results, i.e., the best reliability. As mentioned above, at least 30 specimens of PET melt-spun fibres for each processing condition were conducted 'single fibre tensile test', and then the characteristic Weibull parameter X_0 and Weibull moduli m were obtained by Eqs. (12) and (13). In order to better assess the performance and reliability of the two selected fibres, the average values of the characteristic Weibull parameter X_0 and Weibull modulus m from three virgin PET fibres (fabricated at 270 °C, 2.0 bar and 280 °C, 2.0 bar) were defined as the reference points respectively for two processing temperatures. Then the graphs can be divided into four quadrants using the reference points. Given the reference points, the measured value located in quadrant I shows excellent performance and reliability of specimens; located in quadrant II corresponds to a lower performance but remains an excellent reliability. As interpreted in an analogous manner, the value located in the quadrant III means a lower performance as well as a decreased reliability; located in the quadrant IV means a desirable performance but poor reliability.

In Figs. 21 and 22, m_F, m_T and m_ε over the corresponding characteristic force at break F_0, tenacity T_0 and elongation at break $\varepsilon_{b,0}$ for ***rPET-B*** and ***vPET-1*** fibres were respectively plotted. It can be found that, taking two processing temperatures into consideration, the maximal characteristic force at break F_0, and the characteristic elongation at break $\varepsilon_{b,0}$ for both materials are achieved at 0.5 bar. Accordingly, the main driving force for the performance of force at break and elongation at break should be the take-up pressure, i.e., the draw ratio of fibres. That is, lower take-up pressure leads to a higher force at break and higher elongation at break. Excluding the influence of the diameter of fibres, the maximum of characteristic tenacity T_0 for ***rPET-B*** fibres can be obtained at 270 °C, 3.0 bar, while for ***vPET-1*** fibres is at 280 °C, 3.0 bar. This phenomenon has been well analyzed from the perspective of crystallinity and molecular chain orientation in Sect. 4.5.3.

The highest values of all the Weibull moduli m_F, m_T, m_ε are revealed at 280 °C, 1.0 bar for ***rPET-B*** fibres, as seen in Fig. 21, however, at 280 °C, 0.5 bar for ***vPET-1*** fibres considering the error bar in Fig. 22. Therefore, the possible main driving force (or forces) for the reliability of melt-spun PET fibres in this study could not be identified clearly. The values of Weibull moduli m for each applied parameter are also shown in the charts, which will guide the design on the processing parameters of melt-spun PET fibre in the future.

To analyze the relationship between T_0 and $\varepsilon_{b,0}$, a power law equation was first given by Schubert, Eq. (14):

$$T_0 = K(T) \cdot {\varepsilon_{b,0}}^{\lambda} \tag{14}$$

where $K(T)$ is an adjustable parameter, which relies on the processing temperature during the specific melt spinning process, and presumably also depends on the physical properties of the polymer materials, e.g., molar mass, branching structures, and so forth. A higher value of $K(T)$ denotes a higher tenacity and a higher elongation at break. λ is a material-dependent constant, which yields the same value for the same material.

As presented in Fig. 23a, for the most desirable fibres from ***vPET-1***, the constant value of $\lambda = -1.39$ at both temperatures is obtained by the best fit. The $K(T)$ of ***vPET-1*** at 270 °C, $K(270)$, and at 280 °C, $K(280)$, are calculated respectively to be 7331 and 9907, indicating that for ***vPET-1***, 280 °C is a preferable processing temperature for melt spinning. Both fitting lines in Fig. 23a are transferred to the identical coordinate system in Fig. 23b, in order to provide a benchmark to evaluate

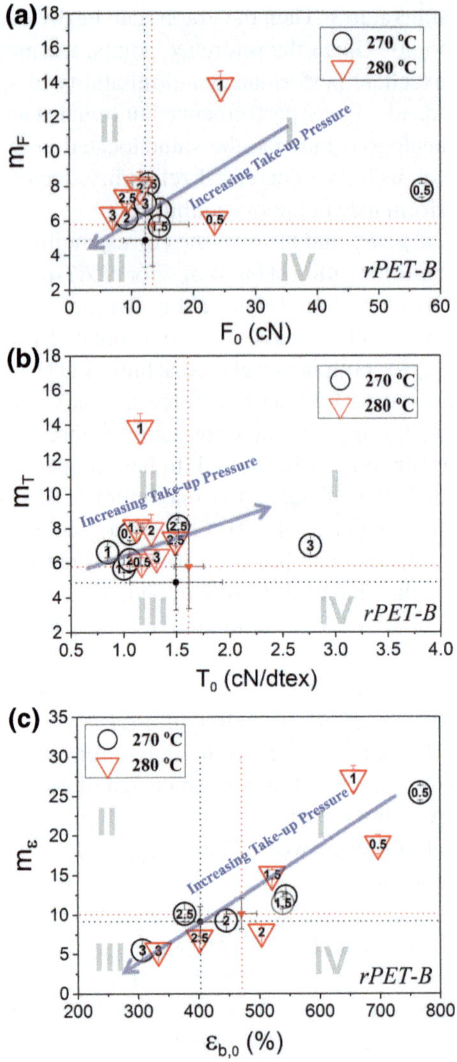

Fig. 21 Performace-reliability charts, plotting Weibull moduli m_F, m_T, m_ε respectively versus the characteristic force at break F_0, characteristic tenacity T_0, and characteristic elongation at break $\varepsilon_{b,0}$ for ***rPET-B*** fibres. The number embedded in the symbol shows the take-up pressure. Defining the average values from three virgin PET (solid black circle: 270 °C, 2.0 bar; solid red triangle: 280 °C, 2.0 bar) as the reference points, quadrants can be constructed as shown

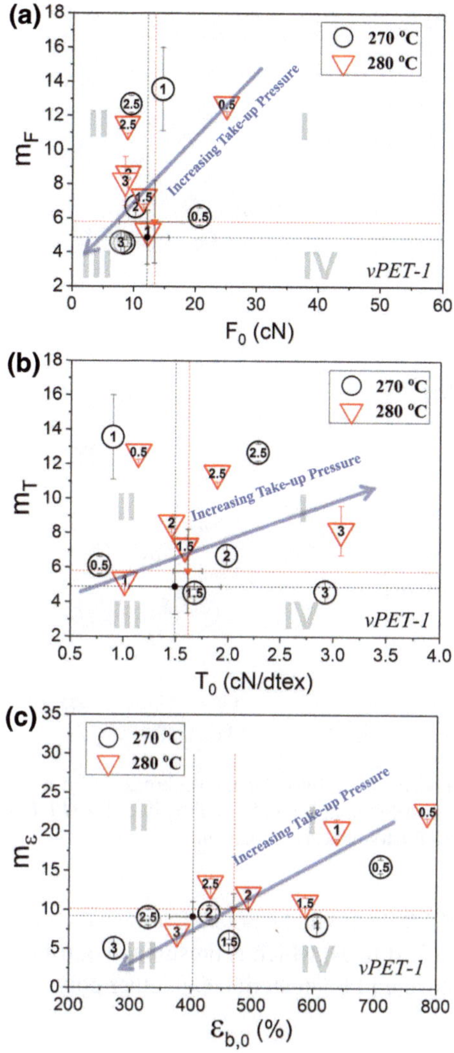

Fig. 22 Performace-reliability charts, plotting Weibull moduli m_F, m_T, m_ε respectively versus the characteristic force at break F_0, characteristic tenacity T_0, and characteristic elongation at break $\varepsilon_{b,0}$ for ***vPET-1*** fibres. The number embedded in the symbol shows the take-up pressure. Defining the average values from three virgin PET (solid black circle: 270 °C, 2.0 bar; solid red triangle: 280 °C, 2.0 bar) as the reference points, quadrants can be constructed as shown

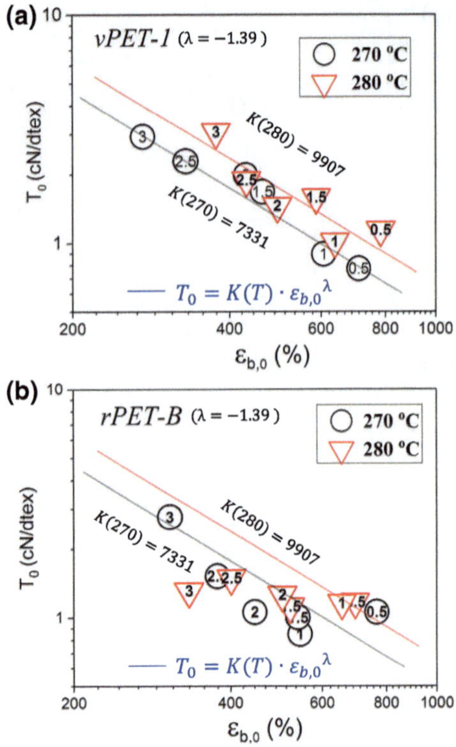

Fig. 23 Characteristic tenacity T_0 versus elongation at break $\varepsilon_{b,0}$ of **a** ***vPET-1*** and **b** ***rPET-B***. The lines for ***vPET-1*** are the best fit using Eq. (14), and the lines for ***rPET-B*** correspond to the same functions as those for ***vPET-1*** to make a comparison

data points from ***rPET-B***. It should be noted that the data points from ***rPET-B*** deviate from the benchmarked lines, which is presumably attributed to the complicated composition of the recycled PET material, e.g., other polymers (see Table 1).

6 Comparison with Reviewed Literature

As discussed above, the PET ***rPET-B*** and ***vPET-1*** can be considered as the optimal materials for the aspirator assisted melt spinning among the five PET materials in this study, due to their best spinnability and mechanical properties. It is also found that the properties of the melt-spun PET fibres from bottle-grade recycled copolymer PET (viz., ***rPET-B***) is similar as that of the fibre-grade virgin homo-polymer PET (viz., ***vPET-1***). Therefore, the properties of ***rPET-B*** fibres manufactured at 270 °C, 3.0 bar and ***vPET-1*** fibres fabricated at 280 °C, 3.0 bar, were chosen to provide a comparison with the data of other melt-spun PET fibres in literature [1, 18, 19, 25].

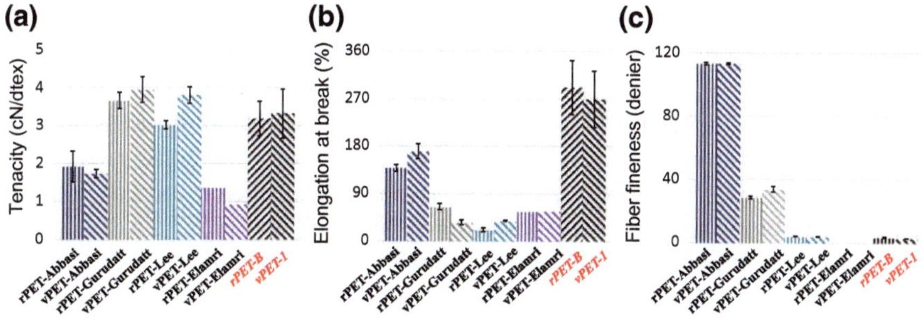

Fig. 24 Three criteria to evaluate the properties of the PET fibres in this study (***rPET-B*** fibres (270 °C, 3.0 bar) and ***vPET-1*** fibres (280 °C, 3.0 bar)) and those from literature [1, 18, 19, 25]: **a** tenacity, **b** elongation at break, and **c** fibre fineness in denier. *Note* No data in denier from Elamri et al. [25] can be collected. Redrawn by authors from Ref. [12]

As presented in Fig. 24, the following three criteria are used to describe the properties of the PET fibres: (a) tenacity, (b) elongation at break and (c) fibre fineness in denier, which denotes the strength of the fibres, the deformation of the fibres and the potential applications for fine filament, respectively.

Though the experimental parameters in the tensile test can bring apparent influence on the results of mechanical properties, as discussed in Sect. 4.5.1, especially the gauge length on the elongation at break. Unfortunately, most of the used gauge length or test speed were not reported in the reviewed literature [1, 18, 19, 25]. Nevertheless, the compared charts in Fig. 24 indicate that both ***rPET-B*** and ***vPET-1*** fibres in this study have achieved a desirable tenacity, an outstanding elongation at break and excellent fibre fineness. It must be pointed out that the fibres in the research of Gurudatt et al. [1] and Lee et al. [18] were stretched at a drawing temperature of 80 °C (near to the T_g of PET), which would increase the crystallinity and orientation of the PET fibres, and thus strengthen them.

Moreover, it has also been demonstrated that the aspirator used in this study can be considered as efficient pilot equipment to mirror the real fibre spinning procedure in the nonwoven industry [62–65].

7 Conclusions

This chapter presented a wide evaluation of melt-spun fibres based on five different recycled and virgin PET materials. Two recycled PET flakes of coloured ***rPET-A*** and clear ***rPET-B*** from waste PET bottles, and three virgin PET materials of bottle-grade ***vPET-2***, fibre-grade ***vPET-1*** and fibre-grade ***vPET-3*** as the reference benchmark were chosen. The capillary rheometer in the laboratory was used to produce PET fibres, which was equipped with an aspirator to provide air flow for aerodynamic stretching of fibres. Two processing temperatures of 270 and 280 °C, combined with six different take-up pressures from 0.5 to 3.0 bar in the aspirator were applied during the fibre spinning process. In this section, an empirical equation was proposed, which could effectively predict the water content of PET materials using the drying temperature and drying time.

The characteristics of the PET raw materials and melt-spun fibres were analyzed comprehensively from many aspects involving thermal behaviour, molar mass characteristic, rheological properties, surface morphology, diameter, thermal stability, tenacity and elongation at break. Accordingly, ***rPET-B*** and ***vPET-1*** with optimal spinnability were proved to be the most suitable PET materials for melt spinning respectively from the recycled and virgin PET. The fibres fabricated from ***rPET-B*** and ***vPET-1*** had a desirable tenacity, an outstanding elongation at break and an excellent fibre fineness. Afterwards, the crystallinity and molecular chain orientation provided an explanation to the improved mechanical properties. Furthermore, a semi-empirical equation with an adjustable parameter and a material-dependent parameter was suggested for the first time relating the tenacity and elongation at break of fibres. In summary, the feasibility has been proved to produce PET fibres from bottle-grade recycled co-polymer PET material, which have similar properties as fibres produced from fibre-grade virgin homo-polymer PET. Generally, the specific information of the investigated PET materials, especially ***rPET-B*** and ***vPET-1*** will provide scientific guidance for melt spinning in the future.

Key Statement

1. Characterization of five raw PET materials: (a) Karl Fischer Titration (water content against drying temperature and drying time, with proposed equation); (b) Thermal properties; (c) Molar mass characterization; (d) Rheological characterization.
2. Characterization of the melt-spun PET fibres: (a) Surface morphology; (b) Diameter distribution of the fibres; (s) Molar mass degradation; (d) Thermal stability; (e) Evaluation of the mechanical properties (tenacity, elongation at break); (f) Crystallinity and (g) Orientation of fibres.
3. Performance-reliability plot using Weibull distribution.
4. Comparison with literature.
5. A semi-empirical equation relating to the tenacity and elongation at break for fibres.

References

1. Gurudatt K, De P, Rakshit AK, Bardhan MK (2003) Spinning fibers from poly (ethylene terephthalate) bottle-grade waste. J Appl Polym Sci 90(13):3536–3545
2. Ghanbari A, Heuzey MC, Carreau PJ, Ton-That MT (2013) A novel approach to control thermal degradation of PET/organoclay nanocomposites and improve clay exfoliation. Polymer 54(4):1361–1369

3. Dimitrov N, Krehula LK, Siročić AP, Hrnjak-Murgić Z (2013) Analysis of recycled PET bottles products by pyrolysis-gas chromatography. Polym Degrad Stab 98(5):972–979
4. Raffa P, Coltelli MB, Savi S, Bianchi S, Castelvetro V (2012) Chain extension and branching of poly (ethylene terephthalate) (PET) with di-and multifunctional epoxy or isocyanate additives: an experimental and modelling study. React Funct Polym 72(1):50–60
5. Shen L, Worrell E, Patel MK (2010) Open-loop recycling: a LCA case study of PET bottle-to-fibre recycling. Resour Conserv Recycl 55(1):34–52
6. Chilton T, Burnley S, Nesaratnam S (2010) A life cycle assessment of the closed-loop recycling and thermal recovery of post-consumer PET. Resour Conserv Recycl 54(12):1241–1249
7. Zhang Y, Guo W, Zhang H, Wu C (2009) Influence of chain extension on the compatibilization and properties of recycled poly (ethylene terephthalate)/linear low density polyethylene blends. Polym Degrad Stab 94(7):1135–1141
8. Awaja F, Pavel D (2005) Recycling of PET. Eur Polymer J 41(7):1453–1477
9. Rajabinejad H, Khajavi R, Rashidi A, Mansouri N, Yazdanshenas ME (2009) Recycling of used bottle grade poly ethyleneterephthalate to nanofibers by melt-electrospinning method. Int J Environ Res 3(4):663–670
10. Zhang H, Wen ZG (2014) The consumption and recycling collection system of PET bottles: a case study of Beijing, China. Waste Manage 34(6):987–998
11. Amano M, Ness B (2004) PET bottle system in Sweden and Japan: an integrated analysis from a life-cycle perspective. Lund, Sweden
12. Qin Y, Qu M, Kaschta J, Schubert DW (2018) Comparing recycled and virgin poly (ethylene terephthalate) melt-spun fibres. Polym Testing 72:364–371
13. Paci M, La Mantia FP (1998) Competition between degradation and chain extension during processing of reclaimed poly (ethylene terephthalate). Polym Degrad Stab 61(3):417–420
14. Edge M, Hayes M, Mohammadian M, Allen NS, Jewitt TS, Brems K, Jones K (1991) Aspects of poly (ethylene terephthalate) degradation for archival life and environmental degradation. Polym Degrad Stab 32(2):131–153
15. Coelho TM, Castro R, Gobbo JA Jr (2011) PET containers in Brazil: opportunities and challenges of a logistics model for post-consumer waste recycling. Resour Conserv Recycl 55(3):291–299
16. Shukla SR, Harad AM (2006) Aminolysis of polyethylene terephthalate waste. Polym Degrad Stab 91(8):1850–1854
17. Sinha V, Patel MR, Patel JV (2010) PET waste management by chemical recycling: a review. J Polym Environ 18(1):8–25
18. Lee JH, Lim KS, Hahm WG, Kim SH (2013) Properties of recycled and virgin poly (ethylene terephthalate) blend fibers. J Appl Polym Sci 128(2):1250–1256
19. Abbasi M, Mojtahedi MRM, Khosroshahi A (2007) Effect of spinning speed on the structure and physical properties of filament yarns produced from used PET bottles. J Appl Polym Sci 103(6):3972–3975
20. Schneiderman DK, Hillmyer MA (2017) 50th anniversary perspective: there is a great future in sustainable polymers. Macromolecules 50(10):3733–3749
21. Jiang Z, Guo Z, Zhang Z, Qi Y, Pu C, Wang Q, Xiao C (2019) Preparation and properties of bottle-recycled polyethylene terephthalate (PET) filaments. Text Res J 89(7):1207–1214
22. Sombatdee S, Amornsakchai T, Saikrasun S (2018) Effects of polylactic acid and rPET minor components on phase evolution, tensile and thermal properties of polyethylene-based composite fibers. Polym Adv Technol 29(3):1123–1137
23. Tapia-Picazo JC, Luna-Bárcenas JG, García-Chávez A, Gonzalez-Nuñez R, Bonilla-Petriciolet A, Alvarez-Castillo A (2014) Polyester fiber production using virgin and recycled PET. Fiber Polym 15(3):547–552
24. Zha AX (2012) Study on production process of PSF from recycled bottle PET flakes for waterproof felt. In: Advanced materials research, vol 441. Trans Tech Publications, pp 764–766
25. Elamri A, Lallam A, Harzallah O, Bencheikh L (2007) Mechanical characterization of melt spun fibers from recycled and virgin PET blends. J Mater Sci 42(19):8271–8278

26. Huang L, Pan L, Inoue T (2007) Effect of solid-state shearing of poly (ethylene terephtalate) on isothermal crystallization and fiber structure formation. J Appl Polym Sci 104(2):787–791
27. Zander N, Gillan M, Sweetser D (2016) Recycled PET nanofibers for water filtration applications. Materials 9(4):247
28. Strain IN, Wu Q, Pourrahimi AM, Hedenqvist MS, Olsson RT, Andersson RL (2015) Electrospinning of recycled PET to generate tough mesomorphic fibre membranes for smoke filtration. J Mater Chem A 3(4):1632–1640
29. Santos RP, Rodrigues BV, Ramires EC, Ruvolo-Filho AC, Frollini E (2015) Bio-based materials from the electrospinning of lignocellulosic sisal fibers and recycled PET. Ind Crops Prod 72:69–76
30. Vo P, Doan H, Kinashi K, Sakai W, Tsutsumi N, Huynh D (2018) Centrifugally spun recycled PET: processing and characterization. Polymers 10(6):680
31. Lim JY, Kim SY (1999) Properties of high-speed spun high molecular weight poly (ethylene terephthalate) filaments. J Appl Polym Sci 71(8):1283–1291
32. Yoon JH, Avci H, Najafi M, Nasri L, Hudson SM, Kotek R (2017) Development of high-tenacity, high-modulus poly (ethylene terephthalate) filaments via a next generation wet-melt-spinning process. Polym Eng Sci 57(2):224–230
33. Jabarin SA, Lofgren EA (1984) Thermal stability of polyethylene terephthalate. Polym Eng Sci 24(13):1056–1063
34. Härth M, Kaschta J, Schubert DW (2015) Rheological study of the reaction kinetics in a poly (ethylene terephthalate) melt. Polym Degrad Stab 120:70–75
35. La Mantia FP, Vinci M (1994) Recycling poly (ethyleneterephthalate). Polym Degrad Stab 45(1):121–125
36. Seo KS, Cloyd JD (1991) Kinetics of hydrolysis and thermal degradation of polyester melts. J Appl Polym Sci 42(3):845–850
37. Darby R, Darby R, Chhabra RP (2001) Chemical engineering fluid mechanics, revised and expanded. CRC Press
38. Qu M, Nilsson F, Schubert D (2018) Effect of filler orientation on the electrical conductivity of carbon Fiber/PMMA composites. Fibers 6(1):3
39. Kong Y, Hay JN (2002) The measurement of the crystallinity of polymers by DSC. Polymer 43(14):3873–3878
40. Starkweather HW Jr, Zoller P, Jones GA (1983) The heat of fusion of poly (ethylene terephthalate). J Polym Sci Polym Phys Ed 21(2):295–299
41. Cruz SA, Zanin M (2006) PET recycling: evaluation of the solid state polymerization process. J Appl Polym Sci 99(5):2117–2123
42. Rizzarelli P, Puglisi C, Montaudo G (2004) Soil burial and enzymatic degradation in solution of aliphatic co-polyesters. Polym Degrad Stab 85(2):855–863
43. Rodríguez-Cabello JC, Santos J, Merino JC, Pastor JM (1996) Thermally induced structural changes in low-shrinkage poly (ethylene terephthalate) fibers. J Polym Sci Part B Polym Phys 34(7):1243–1255
44. Härth M, Kaschta J, Schubert DW (2014) Shear and elongational flow properties of long-chain branched poly (ethylene terephthalates) and correlations to their molecular structure. Macromolecules 47(13):4471–4478
45. Cox WP, Merz EH (1958) Correlation of dynamic and steady flow viscosities. J Polym Sci 28(118):619–622
46. Carreau PJ (1972) Rheological equations from molecular network theories. Trans Soc Rheol 16(1):99–127
47. Ferry JD (1980) Viscoelastic properties of polymers. Wiley
48. Liu C, Li C, Chen P, He J, Fan Q (2004) Influence of long-chain branching on linear viscoelastic flow properties and dielectric relaxation of polycarbonates. Polymer 45(8):2803–2812
49. Wang JS, Porter RS (1995) On the viscosity-temperature behavior of polymer melts. Rheol Acta 34(5):496–503
50. Dealy JM, Read DJ, Larson RG (2018) Structure and rheology of molten polymers: from structure to flow behavior and back again. Carl Hanser Verlag GmbH Co KG

51. Han CD, Jhon MS (1986) Correlations of the first normal stress difference with shear stress and of the storage modulus with loss modulus for homopolymers. J Appl Polym Sci 32(3):3809–3840
52. Kim JK, Jung DS, Kim J (1993) Morphology and rheological behaviour of mixtures of poly (styrene-b-ethylene-co-butylene-styrene) block copolymer and poly (2, 6-dimethyl-1, 4-phenylene ether). Polymer 34(22):4613–4624
53. Han CD, Baek DM, Kim JK (1990) Effect of microdomain structure on the order-disorder transition temperature of polystyrene-block-polyisoprene-block-polystyrene copolymers. Macromolecules 23(2):561–570
54. Jiang Z, Guo Z, Zhang Z, Qi Y, Pu C, Wang Q, Xiao C (2018) Preparation and properties of bottle-recycled polyethylene terephthalate (PET) filaments. Text Res J 0040517518767146
55. Peirce FT (1926) Tensile tests for cotton yarns: "the weakest link" theorems on the strength of long and of composite specimens. J Text Inst 17:T355–T368
56. Litchfield DW, Baird DG (2008) The role of nanoclay in the generation of poly (ethylene terephthalate) fibers with improved modulus and tenacity. Polymer 49(23):5027–5036
57. Fouda IM, Shabana HM (1999) Contribution of interferometry and mechanical parameters in evaluating molecular orientation for drawn polyester. Polym Int 48(7):602–606
58. Baseri S, Karimi M, Morshed M (2013) Effects of tension on mesomorphic transitions and mechanical properties of oriented poly (ethylene terephthalate) fibers under supercritical CO_2 exposure. Polym Bull 70(3):953–969
59. Schubert DW, Kaschta J, Horch RE, Walter BL, Daenicke J (2014) On the failure of silicone breast implants: new insights by mapping the mechanical properties of implant shells. Polym Int 63(2):172–178
60. Lanyi FJ, Kunzelmann P, Schubert DW (2016) Novel chart for representation of material performance and reliability. In: Macromolecular symposia, vol 365, no 1, pp 194–202
61. Bergman B (1984) On the estimation of the Weibull modulus. J Mater Sci Lett 3(8):689–692
62. Chen CH, White JL, Spruiell JE, Goswami BC (1983) Dynamics, air drag, and orientation development in the spunbonding process for nonwoven fabrics. Text Res J 53(1):44–51
63. Abele WP (1998) Lurgi^{TM} nonwoven optimization: a polymer supplier's perspective. Polym Plast Technol Eng 37(3):351–368
64. Jeon BS (2001) Theoretical orientation density function of spunbonded nonwoven fabric. Text Res J 71(6):509–513
65. Hassan MA, Yeom BY, Wilkie A, Pourdeyhimi B, Khan SA (2013) Fabrication of nanofiber meltblown membranes and their filtration properties. J Membr Sci 427:336–344

Test Methods and Identification of Recycled Polyester

P. Senthil Kumar and S. Suganya

Abstract The recycled polyester analysis is a subject matter dealt within in this chapter which draws attention to ascertaining the nature of the fabric, to uncover if they have been unequivocally processed and recycled to what percentage. An updated survey of textile fabrics, processing, recycling and analysis renders knowledge about the molecular weight of the material and purity to meet global demand with breakthrough in the application of recycled polyester. An array of techniques has been engaged including nuclear magnetic resonance (NMR), X-ray fluorescence spectroscopy (XRF), gel permeation/inductively coupled plasma (GPC/ICP) to examine the composition of reclaimed materials or recycled materials from the manufacture of polyester-based materials. Other mechanical/physical testing during the polyester characterization and contamination might attribute to the positive impact on the product while sealing and dispatching is done. In accordance with mechanical testing, the assessment of tensile strength of a material, modulus and strain of polyester is prerequisite. Notably, the route of synthesis substantially affects the properties that in turn to the application of various industry and indoor. Ideal properties of such recycled polyester are broadly ease of designing, corrosion resistant with low density, user-friendly. The resultant is an increased production. Recycling polyester is one of the finest actions to be taken in the dynamic era of the textile industry.

Keywords Polyester recycling · Purity · Molecular weight · Heavy metals · Energy recovery

P. Senthil Kumar (✉) · S. Suganya
Department of Chemical Engineering, SSN College of Engineering, Chennai 603110, India
e-mail: senthilchem8582@gmail.com

P. Senthil Kumar
SSN-Centre for Radiation, Environmental Science and Technology (SSN-CREST), SSN College of Engineering, Chennai 603110, India

S. Suganya
Analytical and Environmental Science Division and Centralized Instrument Facility (AESD & CIF), Central Marine Salt and Chemicals Research Institute (CSMCRI), Bhavnagar 364002, Gujarat, India

S. S. Muthu (ed.), *Recycled Polyester*, Textile Science and Clothing Technology,
https://doi.org/10.1007/978-981-32-9559-9_4

1 Introduction

Textile sector is playing one of the crucial roles in order to implicit world's economy. The nations such as China, India, Italy and German, etc. shares between 10 and 40% of total global textile exports. Among them, India distinguishes itself by the hand-spun and hand-woven textiles and other spectrum is at capital intensive sophisticated mills. While decentralizing power looms/hosiery and knitting sector, the ancient culture and traditions of every nation can be integrated. However, a country's unique product must be suitable for a wide variety of different market segments across the world [1].

The textile industry has spotted a spurt in Foreign Direct Investment (FDI) for the process, particularly in dyeing and printing. The Directorate General of Foreign Trade (DGFT), an automatic route that ensures a number of export promotion policies like incentive rate under the Scheme for the textile sector. Likely, the Amended Technology Up-gradation Fund Scheme (A-TUFS) has emerged in many countries in order to promote textile business and achieve the vision of creating employment and promoting exports through "Make in" with "Zero effect and Zero defect" in manufacturing. A-TUFS is the deciding authority that links the Capital Investment Subsidy (CIS) by 2022. The production of wool specifically is aided by the Integrated Wool Development Programme (IWDP) and which aims to enhance the quality [2].

New skills or innovations are encouraged by the Scheme for Capacity Building in Textile Sector (SCBTS) primarily formed by the Cabinet Committee on Economic Affairs (CCEA). To expand the business, textile parks with a new block-level clusters can be developed through I-ATUFS. It is a web-based claim monitoring and tracking mechanism that buoyed up strong domestic consumption and export demand. However, a fine line between expendable retail sector and consumerism appear to be economically declined due to the entry of international players. Lower disposable income has raised and blatantly affects a huge domestic market.

Solid wastes from textile factories are perturbed with the following disposal techniques of textile goods such as incineration, decomposition, and accumulation. Some of them pose serious threats to ecosystem which is very difficult to screen. Apparel companies are in the demand to secure a certificate permitting that their products have set the international environmental standards and impact-free for both end products and disposable waste. Hence, apparel manufacturers adopt recycling technologies for disposable waste or processed materials to uplift credibility of the goods and services.

Relatively, apparel recycling is a method of reprocessing of pre-owned fabrics, cuts down the cost of sending solid textile waste to landfill; conserves resources; creates awareness among public for participation; adapts to economic market. Textile recycling is hitherto in practice in countries like United Kingdom, Germany, United States and Australia except for China and India. It is to note that Malaysian textile industries cause water pollution apparently. The recycled textile products from Germany and the United Kingdom are surprisingly popular, sustainable and cheered by consumers [3]. For which, they have adopted convenient techniques with more energy sufficient, less cost and talented expertise for public-private-partnerships (PPP).

Polyester is a choice of fabric, man-made synthetic, long-chain polymers composed of esters by 85% weight. It was first introduced in the late 1970s as "a fiber that has a natural ability for almost 68 days wearable and spick and span". A long-chain polymer is the backbone of polyester formed by the esterification condensation of polyfunctional alcohols and acids. The polymer is generally made using polyester yarns or fibers, referring to the most common subclass "polyethylene terephthalate" (PET). The raw material components of PET are generally derived from petrochemicals, with main applications for fiber and packaging production, and a small proportion for film applications. Indeed, polyesters are a medium weight fiber with a density of 1.39 g/cm^3 [4]. As it is heavier than nylon, polyester textile materials are generally manufactured as thin fabrics, since thick polyesters are too heavy. Labels on the polyester fabric present what they are made from and what percentage of natural fabric is mixed with.

The introduction section has dealt with the concept and relevancy of the recycling process. The gaps identified in the business model has well explained to highlight the need for the development of the recycling process and products of textile. Section 2 describes the iterative methodology for the different types of PET recycling. The potential techniques for test and identification of recycled polyester is presented in Sect. 3. In Sect. 4, the proposed 3R sustainability towards the archetype of sustainable business model and their limitation is well elucidated. In a queue, the last section concludes the process methodologies with future development in the recycling business.

2 Textile Recycling/Polyester Recycling

The nonwoven industry has crossed over 50% of polyester production while sharing with the same amount of waste generation per annum. Irrespective of market application and demand for the recycled fabric, the garment industries express their interest to collaborate with academic or research institution to get knowledge for a diverse technical good. Such recycled polymeric granulates and fibers have been proven as effective as conventional insulation materials. Because fibers are recognized as end-of-life textiles, ease of recycling and counting for ensuing products such as clothes and interior monologues for their physical properties.

Germany has advanced/improved technologies for polyester recycling with the help of more than ten partners across the globe. They produce single-material composite systems to be further recyclable as the result of blending various types of textile.

Polyester recycling is a method to reuse or reprocess polyester from the manufacturing process. Recycling of synthetic fur from polyester derivatives is a valuable approach that helps to reduce polymer waste in order to make 'new' high-quality products from existing material. It also makes the textile production chain more sustainable. In order not to destroy polyester's durability (use-life), re-use or recycling potential, the fabric composition, and design have to be made specific. Recycling technologies are broadly classified into primary, secondary, tertiary and quaternary approaches. It is in accordance with the types of waste; the materials how they are sorted out; how stacked based on the condition; consumption of resources like energy and other substances. These approaches can be achieved through mechanical, chemical and biological methods.

Before investment, any recycling practitioner must have the basic idea of what is pre-consumer or post-consumer waste. First, pre-consumer waste is composed of fiber and yarn waste; waste from trials and process tuning from the textile mills (particularly dying); cutting waste from garment manufacturing; and unsold stock from brands, wholesalers and retailers. So, these wastes are recycled into miracle resultant for new indoor applications. Second, post-consumer wastes is the garments or textiles discarded after use because they are haggard, damaged, shrink or gone out of fashion [5]. The clothes thrown away or end up in municipal landfills or incineration plants are categorized here. In an environmental aspect, waste is classified based on the route of disposal. Domestic waste in most of the cases ends up in landfilling rather than burning, creating potential land use with the emission of hazardous substances to soil and groundwater (depending on the way the landfill is controlled). However, a separate collection channel of textile waste is advisable to reuse the virgin raw materials refraining from the low standard waste dump or incineration fire (Fig. 1).

Strategies to recycle is as follows:

- First grade the incoming material into type and color
- Shred and blend with selected fibers (depend upon intended use)
- Clean or bleach the blended mixture
- The end product might be non-woven, used as soft, mattress fillings and furniture padding
- The end product of fibers involves in spinning is much complicated.

Prior to the above strategies, the new making of recycled polyester requires shredding, granulating, melting and then significant extruding. It can replace the use of virgin fibers in each cycle but eventually, the quality of fiber will be decreased and end up in disposal or incineration. This is known as down-cycling.

2.1 *Types of the Recycling Process*

Contemporarily, recycling of polyester takes place in both mechanical and chemical methods (Fig. 2).

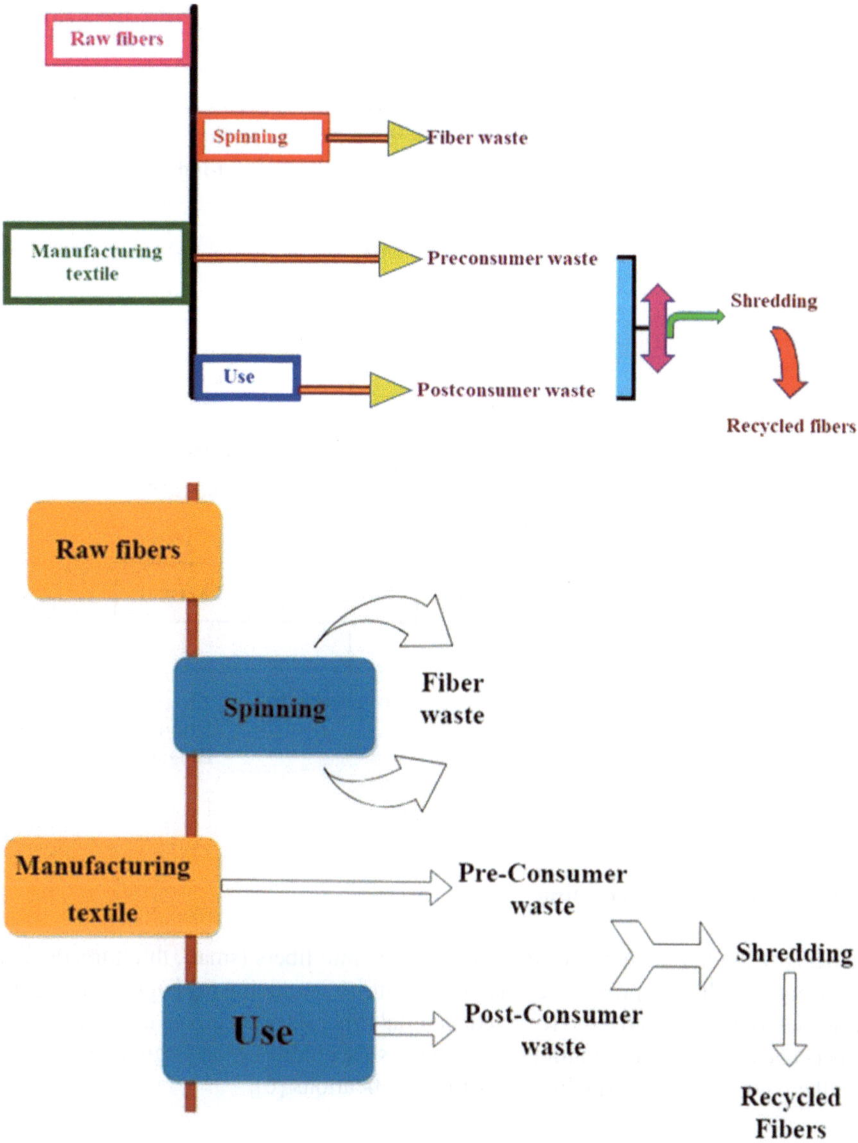

Fig. 1 Types of polyester waste

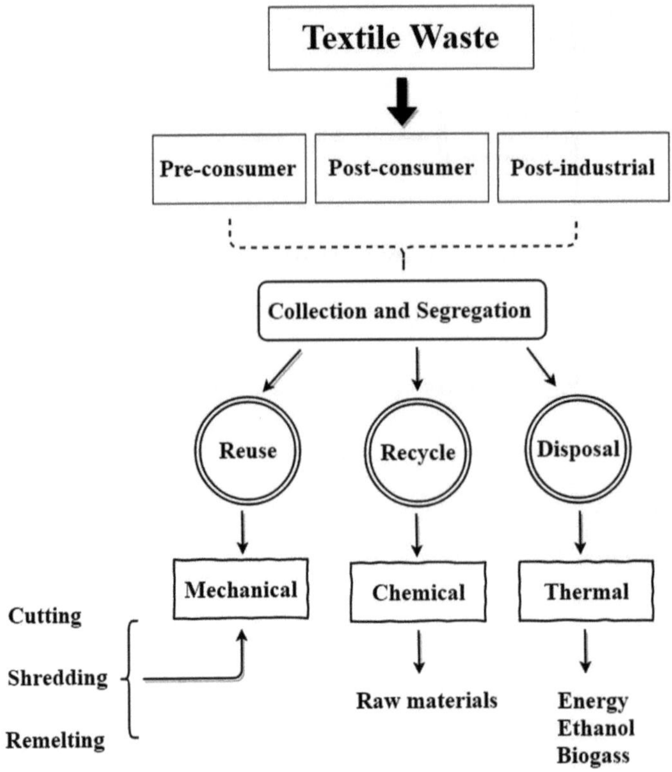

Fig. 2 Types of the recycling process

2.1.1 Mechanical Recycling

It deals with the unraveling of discarded textile into fibers (small, thin threads) and PET bottles. The end product results in films, sheets, strapping packaging and insulation materials. The recovered recycled material (fiber) is often mixed up with original fibers and colors while spinning yarn. Whereas the quality of the virgin fiber is based on the recovery of fiber quality and yarn specifications [6].

2.1.2 Chemical Recycling

It notably deals with the chemical dissolution of the textile waste into its basic chemical building blocks. It results in the new synthetic fiber (recycled polyester). Adequate methods of dissolution are possible through acids, methanol, and glycols. Chemical recycling methods have predominantly confirmed the sustainability of the business and leading to the genesis of virgin stuff. For emergence, the reaction time might be reduced with increased monomer yield by adopting to the following process detailed in the below section. Whereas the reaction takes place at mild conditions [7].

2.1.3 Glycolysis

PET recycling is plausible through glycolysis that taken place in a wide scope of temperatures from 180 to 250 °C. The different catalyst is used to achieve a certain degree of glycolysis. The process of conversion begins in 4–5 h and tailor-made finish is possible in 9 h. The ideal yield of reaction can be obtained in a few seconds with the use of a catalyst. PET glycolysis engages with the use of metal salts, ionic liquids, and hydrotalcite. These catalysts have distinctive properties and executed at assorted environmental condition such as temperature and pressure. There is a breakthrough that enzyme has a degradative biocatalytic activity for PET witnessed by researchers [8]. Therefore, PET degradation by enzymatic catalyst is welcome and given in brief.

Advantages

- Temperature condition
- Depolymerization is faster in lower time
- Highest quality and efficiency are possible with appropriate catalyst.

2.1.4 Metal Salt as Catalyst

Active salts or eco-friendly metal salts are used for the reaction of PET glycolysis. The following salts such as zinc acetate, sodium carbonate and bicarbonate, sodium and potassium sulfate has been explored. The PET glycolysis results in the preliminary color change as such as silver and white; amber (yellow and orange); green and blue for the different substances. On the whole, temperature, ethylene glycol (EG) and PET molar ratio on the reaction prevalently affect the reaction yield. Despite, an increased temperature from 165 to 180 °C has emulated the higher yield. Hence, the optimum temperature can be set as 180 °C. Similarly, increasing EG: PET molar ratio (from 4 to 6) has also improved the reaction [9]. Overall, the reaction is stable and remarkable with respect to the appropriate change in temperature, pressure, and EG: PET ratio (Table 1).

In the note, zinc acetate as a catalyst has been showing the maximum yield as compared with others. The best efficiency was observed within 180–195 °C. The ideal condition of the reaction and the end product can be assessed through toxicity and nonbiodegradability studies. The end product of polymer containing monomer, dimer and trimmer molecules have experienced with the conventional approaches used

Table 1 Chemical structure of various monomers and compounds used in PET analysis

Chemical	Abbreviation used	Chemical structure
Ethylene glycol	EG	$HOCH_2CH_2OH$
Refined tereph-thalic acid	RTA	$HOOC–C_6H_4–COOH$
Dimethyl tereph-thalate	DMT	
Bis hydrox-yethyl tereph-thalate	BHET	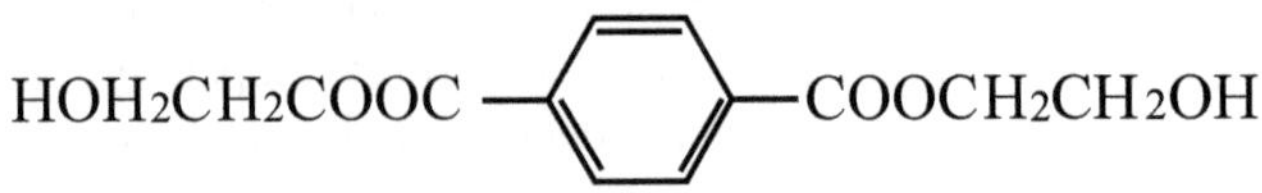
Diethylene glycol	DEG	$HOCH_2CH_2OCH_2CH_2OH$

to separate. However, the aforementioned four biocompatible catalysts (mentioned above) has the best effect in recycled PET via glycolysis as a current trend.

2.1.5 Hydrotalcite

The use of hydrotalcite nanoclays for the PET glycolysis is novel. Hydrotalcite is primarily modified into polycondensation catalyst by various treatments. PET catalysed by hydrotalcite demands more effectiveness in hydrotalcite activity upon decreased with calcination. However, treated catalyst offers higher catalytic activity (>untreated) as the resultant of rehydration. Nevertheless, hydrolytic activity appears fast when the replacement of carbonate anions of hydrotalcite takes place by hydroxide or alkoxide groups. The molar ratio of magnesium to aluminium cations plays a crucial role in the above process. Perhaps, polycondensation results in the expansion of one of the forms of hydrotalcite (sheets and particles) sheets with enhanced activity [10].

The degradation mechanism of hydrotalcite as a catalyst in solvent dimethyl sulfoxide (DMSO) can achieve in 10 min. Likely, the treatment of sodium hydroxide in methanol at 20 °C ends in the precipitation of methanol and ethylene glycol. The reaction temperature can be tuned up to 180 °C at a different time which will eventually provide the best catalytic activity with 98% of conversion. Further separation can be engaged by distillation. After the process, the oligomer can be transferred to dimethyl terephthalate (DMT) and ethylene glycol. The hydrotalcite can be recycled and the oligomer is found to be pure through GC-MC analysis. In context, hydrotalcite acts as a fine recyclable catalyst for the PET degradation.

2.1.6 Ionic Liquids

Ionic liquid is widely used for PET glycolysis with defined atmospheric pressure, temperature and time. The purification of the glycolysis product is simpler, unlike metal acetates. The conversion of PET is possible at 180 °C with 1-butyl-3-methylimidazolium bromide in 8 h [11]. The main advantage is the ease and cost of preparation. Reusability of so far called ionic liquid catalyst is need of the hour. Fe containing magnetic ionic liquid is paid much attention recently for PET glycolysis. This can potentially replace the commercial catalyst with improved selectivity.

2.2 Upcycling and Downcycling

Upcycling is the process in which recycled raw materials or higher-level application are used than the original level (e.g. recycled PET yarn from plastic PET bottles).

Downcycling is the process in which lower level quality material is used (e.g. textile waste to prepare automotive waste).

3 Test Methods of Recycled Polyester

Finished polyester is generally classified into filament, staple, tow and fiberfill.

Filament: Spun polyester is generally called filament since each strand is soften through agitation and spun for long-lasting wear-life. The smooth-surfaced fabric is lengthy.

Staple: PET chips or filaments are broken; cut into short with predefined lengths. They are easier to blend with other fibers and often used in cushion.

Tow: A loosely drawn, continuous and voluminous filaments.

Fiberfill: This type of fiber is combed and stuffed to form voluminous cotton ball.

Among all, filament and staple are the choices of clothing industry [12].

3.1 Polyester Manufacturing

Polyester is formed at 210 °C from dimethyl terephthalate reacting with ethylene glycol. It produces a single molecule known as monomer while raising temperature

Table 2 Intrinsic value of fibers

	Application	Intrinsic viscosity value
Fibre grade	Textiles	0.40–0.70
	Technical	0.72–0.98
Film grade	Biaxially oriented PET film	0.60–0.70
	Thermoforming sheet	0.70–1.00

from 210 to 280 °C. Newly-formed polyester in a long molten-ribbon are subject to drying and cooling to brittle. Such a regular polyester chips are melted at 270 °C to form a syrup-like solution [13]. It is further poured into a spinneret and forced to produce special fibers through tiny holes. Flame retardant can be used in the spinning stage. The mechanical characteristics of a fiber such as strength, tenacity and resilience might fix when stretching force is applied. It is obvious that finished materials are solid, strengthened, textured and twisted.

3.2 The Recovery of Polyester

The intended applications of PET polymer are based on their physical properties, labeled with different grades which ultimately affects recycling. The molecular weight of the polyester is measured in terms of the intrinsic viscosity (IV) that is related to the material's melting point, tensile strength and crystallinity index. PET bottle resin has a higher IV and crystallinity, whereas fiber grade PET textiles and technical applications range from low to high IV values [14] (Table 2).

3.3 Test Methods and Identification of Recycled Polyester

The recycled polyester or PET is identified with density, viscosity and solubility properties using microscopic, mechanical and spectrophotometric measurements. Several traditional methods such as stain and burning test are still in practice. Scanning Electron Microscopy (SEM), Nuclear Magnetic Resonance Spectroscopy (NMR), Fourier Transform Infrared Spectroscopy (FTIR), Raman Spectroscopy, Crystallographic analysis, Thermogravimetric analysis (TGA) and Differential Scanning Calorimetry (DSC) are conventional methods.

3.3.1 Scanning Electron Microscopy

Scanning electron microscopy is a powerful technique used to examine the polyester samples. Microscopic analysis is performed after the conformity of 100% polyester by one of the traditional methods.

Microscopic visualization of the PET Microscopic visualization of the PET or hollow fibers is possible using a polarized light microscope while in combination with SEM. However, most of the synthetic fibers are produced round in shape. The microporous round fiber cross-section can also be envisioned but not preferred. A specialized section of other shapes with clefts like a typical slit, Y and T shape, circle slit and hollow shapes have transpired for reasons, such as performance, comfort, surface defect (pilling), readiness, bulkiness, tactility, processing, etc. Regular PET types have round cross-section unlike others having core shapes [15]. The significant properties of cylindrical fibers such as pilling propensity, bending and stiffness, inelasticity, friction, texture and softness are distinct from a linear cross-sectional shape of fibers. The cores of hollow fibers provide information about the wall thickness/hollowness and thermal comfort of the PET. A compressed thick wall with four channels can be seen and provide thermal protection.

3.3.2 Raman Spectroscopy

Adulterated textile can be examined using vibrational, rotational and low-frequency modes of system using Raman spectroscopy. The r-PET is required to combat fraudulent fiber for which optical spectroscopy like IR spectroscopy is used. The rapid identification offers qualitative and quantitative assessment of polyester samples with reliable well-defined spectral structures. Such a molecular level identification employs scatter light usually monochromatic in the infrared region and often lies on the ultraviolet region. The shift occurs during molecular vibrations or interactions that explicit energy further to provide information about the tested sample [16]. The vibrational identification results in the fingerprint of identified molecular structure that deals with the infrared analysis. Molecular identification includes density and crystallinity with often less polar functional groups.

3.3.3 NMR Spectroscopy

A textile is subjected to NMR spectroscopic analysis via one among following chemical shift, Zeeman Effect or the Knight Shift effect in order to obtain molecular information. A physical state of textile is examined by the resonant frequencies developed during analysis. Similarly, chemical and structural information of solid sample is resulted from topologic and dynamic investigation. Three-dimensional structure of molecules can be seen through low field NMR spectroscopy. Crystallization, spinning and finishing ratio can also be evaluated [17].

3.3.4 X-Ray Diffractometry

Recycled fibers with distinct properties from virgin material is examined using X-ray diffractometry. It is a non-destructive technique rendering knowledge about crystallinity of a fiber, internal structure of a fiber, the shape of the scattering samples, chemical composition and distribution of tested fiber. Crystallinity deals with a wide range of crystalline, semi-crystalline and amorphous based on the spectrum obtained within angle 2θ. The determination of α and β molecular configuration of fiber is done through X-ray scattering on the principle of dynamic theory of diffraction. Small Angle X-ray Scattering (SAXS) probe is available to study between nanoscale to microscale of the fibers whereas scattering angle is $2\theta > 0°$. Likely, the Wide-Angle X-ray Scattering (WAXS) probe is available to study macro structure of fiber whereas scattering angle is $2\theta > 5°$ [18].

3.3.5 Thermogravimetry

As the function of the change in temperature with respect to change in the weight of the polyester sample can be determined using thermogravimetry. Thermal gravimetric analysis (TGA) focusses on the measurements of weight with respect to temperature and temperature range.

Thermoanalytical technique identifies the weight loss of a fiber and at which point the loss is more apparent. The fibrous samples with respect to shrinkage, bending or expansion, moisture content, temperature range which is proportional to thermal stability and fiber strength can be measured [19]. The final products of textile can be further examined using Differential Scanning Calorimetry (for melting behaviour), Thermogravimetry (for evaporation behaviour), Thermo Mechanical Analysis (for shrinkage), Dynamic Mechanical Analysis (for viscoelastic behaviour). Thermal degree of degradation is also be measured for threads, fibers and fabric.

Differential Scanning Calorimetry

TGA analysis provides the information of the mass loss of a fiber with respect to time. The exothermic or endothermic effect with regards to time can be detected from Differential Scanning Calorimetry (DSC) analysis [20]. Since DSC/TGA is less sensitive and DSC curves are also less resolved. Crystallization temperature (Tc) and melting temperature (Tm) can be estimated from the peaks generated by DSC signal [21]. The maximum onset and melting temperature are tabulated [22] in Table 3.

Table 3 DSC temperature profile of fiber

Temperature (°C)	Regular PET
Onset temperature	248.96
Melting temperature	250.28

3.3.6 Fourier Transform Infrared Spectroscopy

Fourier transform infrared spectroscopy (FTIR) is a semi-quantitative technique. A single or two component fibers are quantitatively analysed through obtainment of absorption spectra. The FTIR spectra obtained in the IR region (4000–400 cm^{-1}) particularly in between 2800 and 3300 cm^{-1} attributes to CH_2— groups, 1700–2200 cm^{-1} attributes to carbonyl (C =O) stretching vibration, 1090–1300 cm^{-1} attributes to C–O vibrations. The benzene ring is identified between 600 and 900 cm^{-1}. The intensive peaks between 1400 and 1500 cm^{-1} is distinctive for phenyl and methylene groups in the textile [23]. Both PET and hollow fibers resulted in similar absorption bands.

3.3.7 Gel Permeation Chromatography

Gel permeation chromatography (GPC) or Size exclusion chromatography (SEC) is used in Textile industry. The measurement of molecular weight (MW) and polydispersity index (PDI) [24] can be evaluated from the heterogeneity in molecular mass of fiber. The most apparent mechanical properties like strength, sharpness of thermal transitions [25] can be evaluated from the molecular weight. It also estimates about the aggregation behaviour of the fiber (Table 4).

4 Supply Chain Approach in Textile Business

A fundamental shift from aggressive competition to collaboration imposes sharing is mandatory in the textile business. In every aspect, innovation in business model has become the need of the hour that emphasizes to re-conceptualize the logic and value. Such a redesign can build sustainability into business models for new start-ups. Business models are systematic, playing a main role as a sustainability delivering factor [24, 25] (Fig. 3).

4.1 Research Gaps and Objectives

Textile business is a social enterprise that needs new economy concepts via “sustainable business models”, “closed-loop business models” and Product Service Systems

Table 4 Properties and test results of recycled polyester

Type	Properties	Result
Recycled polyester	Specific gravity	1.22
	Moisture content (%)	0.4
	Tensile strength (N/mm^2)	200–400
	Tensile modulus (GPa)	12.43 ± 2.23
	Length (mm)	20
	Diameter (μm)	25–30
	Aspect ratio	230–300
	Colour	Colourless
	Shape	Straight
	Shrinkage limit (%)	18
	Density ($g.cm^{-3}$)	1.33
	Maximum dry density (kN/m^3)	17–18
	Water absorption (%)	3.3
	Vitreous temperature (°C)	62
	Fusion temperature (°C)	252.8

(PSS). Natural capitalism via circular economy is welcome by many nations. Innovation in technical and organizational sustainable business model have to be brought up. The gap between research and social training in the focussed area, also hence restricts the practical implementation. Sustainability benefits are possible while combining several approaches.

4.2 *Open and Closed-Loop Recycling*

Collection and reprocessing are the two stages involved in recycling. The collected waste from various sources at various points are included in the supply chain. The waste reprocessing might be done either by an open or closed-loop system [26]. In an open-loop system, an indefinite time of recycling is not recommended. Henceforth, exclusion from the utilization loop is possible either reroute to landfill. The facts behind are

(1) Raw material degradation
(2) Non-recyclable product.

In a closed-loop system, it is vice versa that an indefinite recycling takes place without degradation. Though the used product is capable of producing another or same i.e. biodegradable materials.

Fig. 3 Circular economy for the recycled textile fiber

4.3 Sustainability in Recycling

With respect to a rising global population and the utilization of associated resources and environmental impacts, a sustainable textile business is required for the future. The eradication of ecological systems for human welfare and natural capital moves towards today's levels of consumption with unsustainable rate. However, "free" natural assets cannot be practiced in business. Therefore, a sustainable future is a holistic approach with respect to environmental changes parallelly dealing with economic and social change. Sustainable economy deals with the minimal consumption, personal caps on energy, goods, water, etc.; maximizes the institutional and environmental benefit, rather than prioritizing economic growth. It is to be mentioned that the life span of the processed waste or textile materials could be extended technically or emotionally in the phase of their life cycle [27]. The use of higher quality materials in custom-made clothing seems to have a longer life span, intruded shape that can easily get repaired or upgraded. As the separate collection is recommended, used clothing can be organized by a return system.

4.4 Suppliers with a Credible Certificate

There are certain labels regulated by the Control Union Global Recycle Standard (GRS) that certify the use of recycled fibers. The GRS independently verifies claim as the level of recycled material is used in a specific product. The GRS is intended for the strict provision of how waste is handled during the production process; environmental processing criteria and raw material specifications. Certain fabrics like wool, polyester, cotton, and denim, etc., is certified by the GRS, specifying that the percentage of pre-consumer and post-consumer recycled content [28]. Some companies have track and trace system that aims to reveal accurate information about an origin's of item, processing, recycled content and the storage on energy, water, and CO_2. The garment's history is comprised of the end result in a garment tag with a measurement of recycled content.

4.5 Cradle to Cradle

Products with a cradle to cradle certificate must claim product safety towards human and ecosystem, as well as material reutilization through recycling and composting have to prove the same. A four-tired cradle to cradle certification demands basic, silver, gold and platinum levels. Certification is done through a confidentiality agreement.

4.6 Recycled Claim Standard (RCS)

The RCS is as a chain of improvement standard to track recycled materials (virgin) with the help of Materials Traceability Working Group (MTWG) through the supply chain. It comprises of Content Claim Standard (CCS) companies that verifies the recycled content of specific input materials), and additional guidance for implementation of the standard. The CCS Implementation Manual has become mandate for all the textile industries.

The followings are set of selection measures can be adapted to manage or reduce textile waste and increase the use of recycled fiber (polyesters).

(a) To minimize off-cuts and monitor internal reuse and (internal/external) recycling of off-cuts and factory waste;
(b) Trail runs and process start-up and tuning losses, especially in respect to dying;
(c) Recycled Polyester: Polyester is a product of crude oil, just like other plastics such as carbonated drink bottles (PET). Re-cycling these and/or polyester clothes can reduce oil use and energy demands;
(d) The sustainable approach in the design can meet Cradle to cradle approach.

To build up recycling practice in the textile business, the following factors can be enabled. The technology gap between knowledge and implementation is a serious concern.

1. To adopt, stimulate and develop polyester recycling operations across the supply chain (polycotton recycling systems).
2. To automate textile waste sorting and identify fiber-to-fiber methodologies that can incorporate blends i.e. nylon/elastane.
3. To install traceability systems in the existing recyclers which provide information/knowledge of materials and chemicals used in production processes.
4. To ensure quantifying the environmental impacts on safer chemistry and products or implement enzymes, bio-based precursors, etc.
5. To expand technology as not influenced by human exposure, washing/drying, and growth of textile recycling systems.
6. To characterize the recycled products and certify them.
7. To promote awareness among industry and consumers for the sustainable future.

Reclaiming fibers avoids polluting and energy-intensive process that involves in making textile from virgin materials. All recycled fibers score best of class on environmental performance (class A in the MADE-BY fiber benchmark). The Fiber Benchmark (MADE-BY) listed out the most commonly used fibers in the garment industry and their impacts on sustainable alternatives. Twenty-eight fibers are ranked on six parameters: greenhouse gas emissions, human toxicity and eco-toxicity (20% weight), energy, water and land use (13.33% weight). The fibers are placed in Class A to Class E, fibers in 'Unclassified' are not able to be ranked, yet.

4.7 Characteristics of Recycled Polyester

In apparel, polyester fibers are more welcome because they are thermoplastic, or heat-sensitive. Such a high valued polyester can give permanent pleats, decorative shapes and patterns to clothing and flexible for laser-cut into them. Indeed, polyester is harmless, yet, it is prone to static build up. To avoid this, polyester is often blended with more stable fibers, such as cotton. It is referred to polycotton, embodies the benefits of both fabrics; strong, durable, wrinkles resistant and far more breathable than 100% polyester. In general polyester, clothing is bright, slippery, silky to touch and flexible when knitted with other fibers.

- Recycled polyester or PET is durable
- PET is resistant to many chemicals
- PET is resistant to shrinking and stretching
- PET is resistant to wrinkling and abrasions
- PET can blend with lightweight fibers
- PET is ease to subject to dyeing and drying
- The shape of the PET can be retained
- Higher in strength, crease-resistance, and lower water uptake (dries quickly)
- Highly stain-resistant, so ease for cleaning.

In addition, the environmental impacts of polyester become significant. Because wastewater treatment plant has witnessed the presence of substantial amounts of polyester among synthetic microfibers and particles. The condensation agents used were monoethylene glycol (MEG) and purified terephthalic acid (PTA). After which, PET pellets are subject to heating, forming fibers and then spin into filament yarns. 7% of the total polyester production is made from recycled polyester materials [29]. As a synthetic fiber of polyester, can occur to the reaction, despite its limits, it is undeniable that polyester fabric has a huge variety of beneficial uses. A high-grade version or the right blend has come on in leaps and bounds. It is called modern polyester.

5 Conclusion

> More than an image of the material's impact we better to choose to clothe us with recycled and sustainable materials

The textile industry is one of the global industries addressing solid textile waste management and sustainability challenges. Material waste through the supply chain and textile value chain has been a heated argument among producers and suppliers. Creating awareness to the public for the development of sustainable practices is welcome. Looking to the future, collaboration and dialogue among stakeholders are crucial. This chapter has highlighted certain emerging technologies and identification methods in the industry to contribute towards recycled polyester. For future developments in recycling, the practitioners must have textile expertise in assessing and guiding the vision and transition for a circular textile's economy. The closed-loop-system seems to be moving towards a circular economy. The apparel recycled products are the future demand turning on the identification technologies and approaches for material processing from collection to sorting, sorting to processing, and finally producing a new recyclable product.

Creating a demand for new recycled products is critical. The recycled content can be determined in weight percent and certified based on eco-friendliness in reuse with eco-labels. Recycled materials fulfil the following as (1) driven by media scrutiny; (2) producer and consumer demands; (3) sale and cost; and (4) resource scarcity.

The strategy for the textile industry to survive is "invention and innovation" but their supply chain is so complex as time-consuming and difficult in the practical implementation of the schemes and systems. Environmental sustainability has been promoted among the developers of the life cycle assessment of garments. It is critical to address the developing efficient recycled content, sale and closed-loop recycling programs. Despite the competition, hand in hand collaboration can help in progressing the industry.

The development of new methods for the identification of fibers as polyester based or rPET by the chemical solubility tests and DSC analysis for the melting temperature has been discussed. Thus, the melting temperature of the fibers usually to be 250–260 °C. From the FT-IR analysis, the essential active groups can be identified and compared with the regular fiber. Microscopic analyses are essential to study the round cross-section of fibers. From the investigation, Aerelle, Quallofil, and Comforel have only one channel and Hollofil fiber has four channels. Thermal stability can be studied from the thermogravimetric analyzer. Among others, the determination of the morphological difference in hollow rPET fibers is the most useful characterization.

References

1. Bairagi N (2016) Recycling of textiles in India
2. Vadicherla T, Saravanan D, Muthu SS (2014) Polyester recycling—technologies, characterisation, and applications, 150–165
3. Global Fashion Agenda, The Boston Consulting Group (2017) The pulse of the fashion industry 8. http://globalfashionagenda.com/wp-content/uploads/2017/05/Pulse-of-the-Fashion-Industry_2017.pdf
4. Ahmad S, Mulyadi IM, Ibrahim N, Othman AR (2016) The application of recycled textile and innovative spatial design strategies for a recycling centre exhibition space. Procedia Soc Behav Sci 234:525–535
5. Katherine L (2018) Textile recycling technologies, colouring and finishing methods
6. www.ecochiccollection.co.uk/magazine/
7. Khoonkari M, Haghighi AH, Sefidbakht Y, Shekoohi K, Ghaderian A (2015) Chemical recycling of PET wastes with different catalysts. Int J Polym Sci 124524–124535. http://dx.doi.org/10.1155/2015/124524
8. A Report on Recycled Paper (2012) at Natural Resources Defense Council on April 2012
9. Thiele UK, PET recycling concepts for polyester

10. Grigore ME (2017) Methods of recycling, properties and applications of recycled thermoplastic polymers. Recycling 2(24). http://doi.org/10.3390/recycling2040024
11. http://www.fashionpositive.org
12. Ebru B, Buket A, Aslı D, Esen Ö (2012) Potential use of new methods for identification of hollow polyester fibres. Tekstil ve Konfeksiyon 4:317–323
13. http://www.closedlooprecycling.co.uk
14. http://www.patagonia.com/web/us/patagonia.go?assetid=1956
15. Leonas KA, Muthu SS (2017) The use of recycled fibers in fashion and home products. Textiles and clothing sustainability, textile science and clothing technology. http://dx.doi.org/10.1007/978-981-10-2146-6-2
16. Prusty A, Gogoi N, Jassal M et al (2010) Synthesis and characterization of non-fluorinated copolymer emulsions for hydrophobic finishing of cotton textiles. Indian J Fibre Text Res 35: 264–271
17. www.strawberryearth.com
18. http://www.teijinfiber.com/
19. www.tedresearch.net
20. Arthur E (2017) Foundation and circular fibers initiative, a new textiles economy: redesigning fashion's future 91. https://www.ellenmacarthurfoundation.org/assets/downloads/publications/ANew-Textiles-Economy_Full-Report.pdf
21. Abdurrahman T, Özdil N (2015) Effect of recycled PET fibers on the performance properties of knitted fabrics. J Eng Fibers Fabr 10(2)
22. Modulated DSC™ compendium: basic theory & experimental considerations, TA-210, 1–32. http://www.tainst.com
23. European textiles & workwear market: the role of public procurement in making textiles circular. https://www.pianoo.nl/sites/default/files/documents/documents/ecapworkwearreport-maart2017.pdf
24. Bocken NMP, Short SW, Rana P, Evans et al (2014) A literature and practice review to develop sustainable business model archetypes. J Clean Prod 65:42–56
25. Ellen Mc Arthur Foundation and Circular Fibers Initiative (2017) A new textiles economy 48
26. Textile Exchange © 2016
27. http://www.ifm.eng.cam.ac.uk/sustainability/
28. http://www.upcyclingtextiles.net
29. http://www.kategoldsworthy.co.uk/

MIX
Papier aus verantwortungsvollen Quellen
Paper from responsible sources
FSC® C105338
FSC
www.fsc.org

If you have any concerns about our products,
you can contact us on
ProductSafety@springernature.com

In case Publisher is established outside the EU,
the EU authorized representative is:
Springer Nature Customer Service Center GmbH
Europaplatz 3, 69115 Heidelberg, Germany

Printed by Libri Plureos GmbH
in Hamburg, Germany